The Map of My Life

Goro Shimura

# The Map of My Life

Springer

Goro Shimura
Department of Mathematics
Princeton University
506 Fine Hall
Princeton NJ 08544
USA

ISBN: 978-0-387-79714-4        e-ISBN: 978-0-387-79715-1
DOI: 10.1007/978-0-387-79715-1

Library of Congress Control Number: 2008931403

Mathematics Subject Classification (2000): 01-xx

Printed on acid-free paper

springer.com

# TABLE OF CONTENTS

# CHAPTER I

# CHILDHOOD

There are many autobiographies and memoirs without a preface, which I think is normal. In my case, however, the purpose of this book is manifold, and I include many things that may not strictly be called my recollections. Therefore I need to explain why I write them. First of all, I want to write not only about myself but also about my time, or rather, the atmosphere of the time. Thus I include many things that are completely ordinary and known to almost everyone of my generation. I do so because they are so ordinary, they will never be written, and as a consequence, will be forgotten.

Next, I have included my opinions on various historical events and thoughts about human nature, most of which are what I wanted to say at least once, and for which most likely I would not find a platform elsewhere. Such opinions can be found in many autobiographies, and more of them appear in this book.

In the latter part of the book, I write about mathematics, and inevitably I use technical terms. I would advise the reader who is not familiar with these terms simply to read through the passages in question without trying too hard to understand them. After all, this book is not about mathematics, but about how my life is related to the development of mathematics, and I believe that to understand that interaction, it is not necessary to have any special knowledge of mathematics.

G. Shimura, *The Map of My Life,* doi: 10.1007/978-0-387-79715-1_1,
© Springer Science+Business Media, LLC 2008

## 1. The World of the Kiri-ezu

My ancestors were samurai retainers of a feudal lord in the Edo period (1603–1867). Tokyo was called Edo then. Each lord was the ruler of his fief, but was required to come to Edo and reside there for several months or so. Then he would return to his fief and this was repeated. Of course some retainers and their families followed him, but some of them had permanent residence in Edo. Those retainers were called *jofu,* which roughly means "constantly in the capital." My ancestors were such jofu for about seven generations. Even their tombs were in Edo. My great-grandfather was the last one who served the lord. When I was a child, I thought anything related to Edo was something in the days long past, but the period between 1867 and my birth is more than ten years shorter than my age now. Indeed, the streets of Tokyo in my childhood were, albeit wider and paved, in the same position as in the Edo period. I think they remained so even until around 1970.

None of my forefathers made a name in history. There are only two historical documents in which the name of one of my ancestors appears. In the mid-nineteenth century, a few publishers in Edo sold maps of the city, about thirty sheets for the whole city. Each sheet, about eighteen inches square covering about one square mile, indicates the plots occupied by feudal lords, and the names of their jofu-retainers are written in many small rectangular plots adjacent to the plot of their lord. Some Buddhist temples and Shinto shrines are also shown on the map, but the quarters occupied by the working class are squeezed into disproportionately small areas.

Shimura Kosanji, the name of my great-great-great-grandfather, is one of the hundreds of such names found on two maps published in 1851 and 1854. The two maps are nearly identical, and cover a district called Ushigome, a part of Shinjuku Ward of the present-day Tokyo. In fact, I spent much of my childhood within the same district, and so I walked on the same roads and visited the same Shinto shrines as Kosanji did. The maps indicate also the Buddhist temple on whose grounds was the cemetery where my ancestors were buried. Most of them were born and died in Edo, which was the case with my grandfather too. The maps of this type are called *Edo kiri-ezu.* Hereafter I refer to the particular one showing the house of Kosanji as *the Ushigome Kiri-ezu,* or simply *the Kiri-ezu.*

Apparently Kosanji was not of high rank, because, if he were, his descendants would have remembered him to be so, and would have boasted about the fact, but there is no such thing. However, Kosanji was a diligent man. He left many small notebooks in which he recorded various facts concerning his service to the lord. For example, the lord would give a banquet in honor of some important person; Kosanji would be in charge of it, and would record the menu in his meticulous hand. Also, he learned judo (jiu-jitsu) and kendo (kenjitsu, Japanese fencing) with a teacher, and reached the highest level in both subjects, as evidenced by the fact that the teacher gave him two certificates of the most formal kind.

Each certificate is a scroll, whose main part consists of a long list of various martial techniques. Then there is a pedigree showing who taught whom forming a long line that ends with the name Shimura Kosanji. There is also a warning that everything must be kept secret; if anything is disclosed to an unauthorized person, then the violator will be punished by Marishisonten (Marici, a Hindu god of war). I remember especially the item on the judo scroll concerning how to tie a criminal with a cord. The note below it says that "it is a standard rule that the cost of the cord is 333 mon." A mon roughly means a penny. Perhaps it meant that the teacher would give a cord to the disciple in addition to the scroll, and the latter would pay the former 333 mon.

I think the scrolls are authentic, and I am certain that Kosanji and his teacher took them seriously. For my family members, however, they were objects of mere curiosity. I was very disappointed when I looked at them for the first time, as it looked like the table of contents of a book without text. But of course one could not expect anything other than that. The teachers of other activities such as flower arrangement, music, calligraphy, and dancing issued certificates in those days, and do so even today. But I don't think that any such certificate was a scroll. Why scrolls? I think they needed a long piece of paper to list all the items that were taught. Then it would have been natural to make a scroll, though of course a scroll would have made the certificate look grander and more valuable.

In the Edo period there was a school of mathematics, whose originator was Seki Takakazu. He and his successors were able to develop the theory of algebraic equations and even infinite series; they almost invented differential and integral calculus in their own way, but did not reach the refined level of Western mathematicians. In any case, what they had achieved was quite sophisticated, and I'm sure they gave out certificates to their students, and I even think that some teachers warned their students with possible punishment by some god, if not Marishisonten, but I doubt that the certificates were scrolls.

As for my great-grandfather, Issei, the family legend says the following. At the time of the Meiji Restoration (1868), he naturally became unemployed, but somebody connected to the well-known aristocratic Konoe family gave him a chance of serving one of the Konoes. Konoe is the family name of court nobles of the highest rank, belonging to the Fujiwara clan. However, Issei decided against the idea, saying that "A loyal samurai does not have two masters."

I must emphasize that none of his family were proud of him for saying so. On the contrary, everybody resented him. Apparently Issei was a lazy man. My grandfather, Kintaro, referring to his classmate who later became a well-known professor at the University of Tokyo, used to say, "I was better than he in elementary school. Therefore I would have been more successful in my career, if my father had been employed and had let me get through college," which is of course a standard line uttered by any frustrated person. After giving birth to her first son Kintaro, Issei's first wife died. Issei then had a second wife, who came from a family richer and of higher rank than the Shimuras.

It is said, however, that she was not very smart, and too carefree. She would say, "Today I am afraid I am short of money. I think I'd better sell one of my silk robes." In fact, she had brought many silk robes when she married Issei. In those days, a fishmonger, besides being a shopkeeper, would also visit the home of every regular customer, carrying fish in an oval tub. After selling a robe, she would say, "That large bonito in the tub looks fresh; I must buy it." Then she would buy it. Such a woman was she, but no family member resented her. On the contrary, she was gentle and liked by everybody. Kintaro, especially, had fond memories of her as a very kind stepmother.

In the Edo period and even after the Meiji Restoration any marriage of the samurai class was normally arranged between families of the same fief. That was so for the wife of Kintaro, but this time she came from one of the poorest families in the fief. Also, she was not very kind to her daughter-in-law, my mother. It causes me deep regret that I am related by blood to the lazy samurai but not to the kindly woman who exchanged silk robes for fresh fish.

My father inherited various things, including the scrolls. There were several swords, a spear, a halberd, a matchlock, and other old-fashioned weapons, as well as some tea-bowls and tea utensils for the tea ceremony, owned by Issei's second wife. All these were kept in a large oblong wooden box. My father was employed by a bank, and was often transferred from one branch to the other. Therefore I guess he carried such relics of the feudal age whenever he moved, and I feel sorry for him. My brother still has the scrolls but all else were destroyed by the American bombing during the war.

My grandfather on my mother's side apparently had no regular occupation. He had no son but four daughters, of whom my mother was the third. She was neither proud nor ashamed of him; she only mentioned tidbits of his easy-going life-style, but I gathered from what she told me that he was lazy too, and so I inherited another lazy fellow's blood. She was born in the western rural part of Nagoya, which had belonged to the fief of the lord of my ancestors. Therefore my father married a woman of the same fief as his father

and grandfather did. I suspect that her reticence about her father
had its roots in her ambivalence to her rural upbringing, because
her later life as an adult was completely urban. Here is one of
her favorite stories about her experiences in elementary school.
One day her teacher made her stand in the corner. She thought
it unjust, stomped out of the school, and ran to her house, while
her classmates noisily cheered her on from the windows. Before
narrating another of her stories, I first have to explain a folk tale
every Japanese knows, which is about an injured sparrow and an
old man who saves her.

After letting the recovered sparrow return home, one day he
tried to visit her, shouting loudly, "Sparrow, sparrow, where is
your home?" He finally found her living in a bamboo grove, and
was entertained by her family. When he bade them farewell, they
brought out two wicker baskets, one large and the other small.
Then they let the old man pick whichever he liked. Being modest,
he chose the small one. Arriving at his home, he found a great
quantity of treasures in the basket. One of his neighbors, a mean
old woman, envious of his luck, injured a sparrow, nursed it, let it
go, and went to the same place in the bamboo grove. Being greedy,
she demanded the large basket, and got it. As it was heavy, she
took a rest on the road. She was anxious to see inside, and opened
it. Out came an army of evil monsters, who devoured the wicked
woman.

At my mother's elementary school, a music class consisted of
the pupils singing a simple song, accompanied by an organ played
by an old teacher. The song goes like this:

> Sparrow, sparrow, where is your home?
> Chi-chi-chi, chi-chi-chi, this way in the bamboo grove.

According to my mother, this was customary not only in the first
grade, but also in *every* grade; the pupils were always singing,
"Chi-chi-chi, chi-chi-chi, ... ." I am inclined to believe her, as she
was not very talented in creating a story.

However, I hear that most professors in law or social science
teaching at Japanese universities simply read their notebooks, and

let the students copy what they read. Also they use the same notebooks year after year, and therefore they are singing their own "Chi-chi-chi, chi-chi-chi, . . . ." The maxim "A university professor is the next easiest profession after a beggar" really rings true.

I was born on February 23, 1930 in Hamamatsu, a city 135 miles west of Tokyo, whose population was more than 200,000 at that time. I have no recollection of my life there. According to what my mother told me, my family moved from one house to another in the city. I visited the old place after moving, and came back crying and shouting, "I saw a caaaaat!" She amused herself by repeating this story to me, and even told it to my wife. I suppose her point was, "He, who has such a solemn face now, was once like that." I don't find it interesting, as I have no recollection of it. It would have been more interesting if I had seen a grinning caaaaat.

What I remember about my childhood begins in March, 1933, when my family moved to Tokyo. We lived in an old-fashioned one-story house situated very close to where my ancestors Kosanji and Issei lived, but that was a coincidence.

There was a dense network of streetcar lines operated by the municipal government of Tokyo. One-way fare for an adult was seven sen, irrespective of the distance; a sen was equivalent to half an American penny then. Transfer tickets and various kinds of discount were available, so the system was a convenient and practical way of transportation, comparable to, and possibly more so than, the present subway system in Tokyo; the only drawback was its slowness. The first subway line was built in 1927, and in 1939 it became the present Ginza line connecting Asakusa with Shibuya. It was the only subway line in Tokyo until 1959, when another line was built. In addition, trains ran on state-owned railroads, the Yama-no-te and Chu-oh lines, which are the same as those of today.

In April 1936 I was enrolled in the elementary school situated next to a Shinto shrine which was shown almost at the center of the Kiri-ezu. Japan was becoming politically unstable in the 1930s and there was a well-known failed coup d'état on February 26 that year, in which two ministers and a high army official were killed by soldiers. But the daily life of an ordinary citizen was not much affected. In fact, I remember the five years after my coming to Tokyo as one of the happiest periods in my life.

In 1938 my family moved to a house in Nishi-Ohkubo, also in Shinjuku Ward, which was about a fifteen-minute walk from the old place. I still went to the same elementary school until the fourth grade, but changed to the school in Nishi-Ohkubo, and completed the fifth and sixth grades there.

Speaking about those happy five years, I think the atmosphere of a large shopping district like Shinjuku was not much different from today, and I remember a certain feeling of affluence of the area, though of course there were poor quarters elsewhere. Besides, the streets of Tokyo were not beautiful, if not dirty. The novelist Lafcadio Hearn (1850–1904), who became a Japanese citizen with the Japanese name Koizumi Yakumo, thought so, and disliked Tokyo not only for that reason but also because there were some English and American men in the city with whom he did not want to be associated, as he always had the feeling of being an outsider. However, he eventually came to Tokyo, and lived in a house in the Ushigome district near the Buddhist temple called Kobudera, which is clearly shown on the Ushigome Kiri-ezu. After several years, he and his family moved to a new house in Nishi-Ohkubo, which was very close to my elementary school, and he died there.

The Kiri-ezu also shows a small shrine dedicated to Benten (Sarasvati, a Hindu goddess of water), which is usually called Nuke-Benten for some reason, and another shrine called Nishimuki Tenjin, dedicated to a well-known poet-scholar who died in 903, which once occupied a relatively large domain for such a local shrine. These shrines were just a few minutes walk from my house and also Kosanji's. The streetcar line No. 13 ran from east to west on the road that divided the Kiri-ezu in two. One of the stops of this line was just in front of Nuke-Benten, where my family boarded the streetcar whenever we went to Shinjuku. There was (and still is) a slope going down westward from the Benten shrine, lined with many stores, and my family usually shopped for everyday goods there. There were also tradesmen taking orders at their regular customers' home for commodities such as rice, soy sauce, oil, vinegar, sake, charcoal, and firewood. At the end of the slope, there was a middle school, which, according to rumors at the time, was for boys of rich families who were not smart.

Halfway down the slope there was a Chinese restaurant called Ko-Ran. Whenever I walked by that place, I smelled something that suggested delicious cuisines, and so I had a vague desire to eat in that restaurant someday, but that never happened, though my family had dishes supplied by them when we had guests. To do honor to my parents, I must note that they later brought me to some good restaurants, perhaps better than Ko-Ran.

There were some places of business of the type we do not find these days. Next to the shrine of Benten was a shop that made bows and arrows; a half dozen men, sitting on a wooden floor, would be shaving bamboo sticks. Just opposite was a small printing press, which would make loud noises with moving printing machines. There was also a small place which we children called *Hebi-ya* translated as the "snake shop." It displayed a viper in a small glass bottle placed in the window. I never saw any customer going in there.

I also remember an advertisement on a wall for a dressmaker called Midori-ya. In those days, practically all housewives were in kimono, except in the summer, when they wore simple clothes in the Western style. I never saw any pupil in elementary school in kimono. Middle schools were not coeducational. The boys and girls from seventh through eleventh grade wore uniforms stipulated by each school. The girls even wore raincoats designed especially for the school. As for Midori-ya, I never knew where it was; I am not even sure whether it was really a dressmaker, as the advertisement simply said "Dresses Midori-ya". However, as there was not much demand for Western-style women's clothing, it is inconceivable that it was a retail shop for such goods. Japanese women in

Western clothes became common, or rather, standard only after the war.

On the opposite side of the Midori-ya advertisement was a cheap restaurant, which had a sign for shaved ice in the summer. The front of the place was always open, so that one could see a few men, who looked like laborers, sitting on the chairs. After a few years, that house along with adjacent buildings was replaced by some houses with colorful roof tiles. They were of the type that was called *bunka jutaku* in Japanese, which may be translated as "up-to-date residences." Walking northward from my house a few minutes, I once observed a few girls standing on the rooftop of a two-story Western-style house painted in cream, which was later destroyed by a fire one winter night.

For a pre-school boy like me, the snake shop, Midori-ya, the cheap restaurant, bunka jutaku, the Western-style house in cream, and the girls on its rooftop were unknowable objects of my curiosity, which was never satisfied.

Two houses and ours shared a courtyard behind a gate with a pair of wooden doors facing the street. The style of our house was out of date, but had some interesting points which I describe later. The courtyard had more than adequate space for children to play. At night the gate would be bolted, but there was a side door which I think was not locked.

The house in front of the gate was occupied by a retired army general, and had a pair of gateposts made of granite. The family's New Year decoration pines were the tallest in the neighborhood. I thought he was a very old man, but he was perhaps in his late sixties. His son was an army officer killed in a battle somewhere some years ago, survived by his wife and son. They lived in a small house built on the same property, separate from the main building. The boy was my classmate in elementary school and also a playmate even in pre-school days. I often had snacks with him in his house. We two, together with his mother and grandfather, would comment on the fish in the tub brought by a fishmonger to their place. There was nothing extraordinary or mysterious about this family, as I could see practically everything there.

I had a memorable experience with their northside neighbor, however. One day at the age of four or five, I was playing with two children of the same age in the courtyard in front of my house, when a young woman opened a door on the wall on the other side of the street, smiled at us, and invited us into the garden of her house. I was later told that she was a newlywed, and was living with her husband and parents-in-law. She was pretty, about twenty years old, and wore a simple but elegant kimono. I don't remember what we did then; perhaps she gave us cookies. It was my vague impression that somebody advised her to let us leave, though she and we children wanted to be together much longer, and we were naturally disappointed. The memory of this visit remained a puzzle, if I may call it so, which was never solved. It may have been simply that she saw a few innocent and sweet boys, and spontaneously wanted to play with them. But the door was never opened again for me, and she and the house became another unknowable existence.

There are some ordinary goods that are commonly available today, but not in those days. Let me mention some of them. Electric washers and vacuum cleaners are such appliances; electric refrigerators existed, but not in ordinary households, which used iceboxes, with chunks of ice delivered from an ice shop. The state-owned radio broadcasting system started in 1925 but there was no television. Some of the most popular radio programs were broadcasts of intercollegiate baseball games. But they were interrupted occasionally by stock quotations, which everybody detested. Professional baseball was played for the first time in 1936 among the seven teams established then, but they gained the same degree of popularity as intercollegiate ones only after the war.

There were no electric rice cookers. Every household in Tokyo had gas ranges, and some families used them for cooking rice, but there were many who heated the pot on a kitchen stove by burning firewood, and my family was one of them. This was so until around

1938. The iron pot was made for the single purpose of cooking rice, and had a heavy wooden lid. The rice cooker works even in one's sleep, but one cannot burn firewood while sleeping. Therefore my mother would rise at 5:30 in the morning to start the fire in the stove. In those days I used to sleep next to her, and I would also rise at the same time, and help her tend the burning wood in the stove, and do a few other things necessary for our breakfast. My brother and sisters would still be in deep sleep. Later I read the biographies of several dutiful sons born into poor families who became famous men. They gave me an inferiority complex, as I thought I was not as good as they, but I consoled myself by thinking that, "Well, I at least tended the kitchen stove." At that time I would go to bed before eight in the evening.

Some Western-style buildings built relatively recently, such as colleges, hospitals, and government facilities, had radiators. That was so for some elementary schools that were rebuilt after the Great Earthquake of 1923. But the elementary and middle schools I attended had not been damaged by the earthquake, and as a consequence were heated by coal stoves. Also, most families, except for very rich ones, warmed themselves with a charcoal fire in a brazier. It is well-known that Lafcadio Hearn complained bitterly about the coldness of Japanese houses in the winter, in both Matsue and Tokyo. More than thirty years later we were still shivering from the cold, as he had been. I would be shivering even some twenty more years later, to which I turn in a later chapter.

Few homes had telephones. Many households had maids but not telephones. It is quite possible that maids outnumbered telephones in Tokyo households in the early 1930s. Car ownership in those days meant owning a chauffeured automobile. Taxis were abundant; it was easy to hail one on any busy street. There was a fifty-yen silver coin, which was roughly equivalent to an American quarter at that time, and which was the base price for a taxi ride, and one could even bargain for a discount.

There were many department stores in Tokyo, which did not look much different from those of today; the elevators and

such gardens, was an exception, as it was occupied by the army until the end of the war. At present, only a small portion of it, called Toyama Park, is open to the public, the remainder being sites for schools, apartments, and government buildings.

In Toyama Villa there was, and still is, a manmade hill called Hakoneyama, about 45 meters above sea level, which was the highest point in the old city of Tokyo in the 1930s. There was also a Shinto shrine called "Ana Hachiman," abutting the north side of the garden; *ana* means hole or cave. There was indeed a small cave in the precincts of the shrine. A magazine for boys at that time carried a serial adventure novel about a masked samurai who reached Edo Castle from the cave through an underground passage. Thus, I might say, my childhood was connected to Edo in that fashion. One of the best-known authors of such novels published by the magazine was a teacher at my elementary school, but he quit teaching when I was a second grader.

As I said earlier, my family moved to a house in Ohkubo, which was two-storied in contrast to the previous one-story house. That made my life somewhat more interesting, as I was able to move from one upstairs room to the other by walking on the tiled roof, and even to a room downstairs by climbing down a tree. I am certain that my parents were unaware of such acts of mine. Also I was able to see Mt. Fuji clearly from a window on the western side. The house was well-made and relatively new, but lacked the character of the old house; the garden was far smaller. It is said that a rich man kept his mistress (or rather, one of his mistresses) there. That may have been true, and her secret lover might have carried out the same athletic performance on the roof as I did. But I was not old enough to imagine such; I was merely at the age of being disappointed by the fact that the persimmon tree of the new house was of the astringent type.

I am the youngest child with three sisters and a brother, and naturally was brought up indulgently. However, I don't think that affected me in any negative way. I often observed the self-centeredness of those who were the eldest son or the eldest daughter in Japanese families. They didn't even know how self-centered they were.

Those five years after my coming to Tokyo formed one of the happiest periods in my life as I already said, possibly reinforced by my indulgent upbringing, even though the political situation in Japan was unstable. In any case I keep the happy memory of these five years in contrast to the next eight years, which gradually made the lives of Japanese people gloomier and darker. This happy memory occasionally makes me wonder: Suppose I were brought up without the war; then I may have become a lazy and feeble person like someone who would have run away with a dancer on the Odakyu line and would have needed somebody for the task of cleaning up the mess caused by his elopement. Although I was only fifteen when the war ended, it is certain that my experiences of various kinds during the war such as the American air raids taught me many invaluable lessons. One says, "Undergoing hardships makes you tough and wise," which may be true, but I think hardships sometimes create warped personalities too. We never know. I write more about the air raids in later chapters.

When I was four or five years old, my mother and I would be left alone in the house, as my father would leave for work and my brother and sisters would go to school. Sometimes my mother would go shopping, leaving me alone, and even without locking up, but nothing happened. I once visited a neighbor, and played with a girl there of the same age. We colored pictures in the coloring books with her crayons. She had a fair number of crayons, but she would let me use only very short ones of dull colors; she would keep the long and the bright colored ones for herself. Simply put, she was mean, and that was my first experience with a mean person in my life. Her behavior was of course negligible, but I never forgot it. I met various mean people later, and each incident of someone's meanness left a strong impression on me, perhaps because I was lucky enough to meet relatively few of them. I describe another example later.

I had a strange experience around the same time. I was playing with children of my age in another house. There was an ordinary wooden staircase going up to the second floor. What we did was jump down from the third or fourth step to the landing, where a futon mat had been placed. Some of us jumped from the fifth or even sixth step. At some point, I said, "Well, I will jump from the top," and I did. My feet reached the mat safely, but I fell on my buttocks, so that the tip of my coccyx hit something, and I was unable to breathe for a few seconds. However, I soon recovered and there were no ill aftereffects. I later wondered if I really jumped from the top, but that is what I remember. I was very skinny and I didn't weigh much, which helped me, but clearly my act was extremely dangerous. Naturally I did not tell this incident to anybody in my family. Later I acted recklessly on several more occasions, and perhaps I did so more often than the average person. But that's part of my character, and this was the very first of such acts of mine.

I was not a frail child, as I engaged myself in such an act, but I could not be called robust. Indeed, in winter I caught cold easily and was literally a snotty boy. I once had a tumor in one of my legs, and had it removed at a hospital situated close to the center of the Kiri-ezu; I remember very well the tenderness of the nurses there. I also suffered from whooping cough, and was treated with an inspirator for many days. None of my ailments was serious, however, and I managed to pass my pre-school years happily for the most part. I was in good health after entering elementary school, certainly better than the earlier years.

In those days every elementary school pupil carried a *randoseru* (which came from the Dutch word *ransel*), a knapsack made of leather. So, a few days before entering school, my mother took me to a department store in Shinjuku to buy one. At that time the standard color of a knapsack was black for a boy, and red or pink for a girl. There were naturally plenty of knapsacks in the store, and she asked me which one I liked. Somehow I was attracted to one in light brown that had a soft feel, and picked it. She smiled and bought it without comment.

On the first day of school I discovered that practically all new pupils had black, red, or pink ones; mine, with the possible exception of one other, was the only one in a different color. Why do I note this? There are people who like to stand out; there are also those who intentionally do things differently from others. I do not belong to either group. I just do things my own way, and whether it is standard or fashionable is not my concern. Of course I will avoid going against common sense or being offensive to other people, but otherwise I never try to conform myself to others; nor do I try to be a nonconformist. I only try to be honest to myself, and as a consequence I may look like a nonconformist, but that's not intentional. I just mentioned an early example.

The knapsack I chose turned out to be of good quality, and it lasted all six years in elementary school. Also, its color made the task of finding it very easy. There was one more good thing: nobody made fun of me for having such a knapsack.

My school life started enjoyably except for one minor thing. I belonged to the shortest group of my class, and a member of that group, whom I call K here, played various tricks on me, by hitting or pinching me. He would flee after each act. Nothing was serious, and for a while I let him do as he liked. But it became so annoying that one day I consulted my brother and parents. I don't remember exactly what they said, but their advice was, in brief, I'd better teach him a lesson. There was an unwritten code of honor that such a matter must be settled among us pupils, and we must not tell the teacher. So the next time when K did something and tried to flee, I caught him, and using one of my sumo tricks, threw him to the ground. I then sat astride him, and made him swear that he would never do such things again. It worked perfectly, and since then there was nothing awkward between K and me. In fact, I had practiced that trick with my brother a few times in anticipation. That was the only time in my life I committed such a physical act on somebody. I guess K was very surprised. It was as if he pulled the tail of a cat, and was scratched. But then again, the cat would not have sat on him. Possibly he wanted to attract my attention, and did not know how to do it.

K was very good at composition. There was a movement to emphasize the importance of composition in elementary school education. This started around 1915, and all elementary schools in Tokyo up to 1940 were under the influence of this trend. In that period there were seven or eight elementary schools in Ushigome Ward, and the best compositions of the pupils of those schools were printed in a booklet every year and distributed to all the pupils. The types of compositions particularly encouraged by the teachers were those that described everyday life of the pupil's family, which were called "life composition." Naturally the life of a working class family made a better story, and a collection of compositions of that type by a girl named Masako Toyoda, published in 1937, had become a best seller. There would be a revival of this movement around 1951, but that is another story. K's family was upper-middle, but he still had a talent of weaving a story smoothly.

One day in the second grade, the teacher asked me to make a neat copy of a rather long composition by K. That was because, I believe, the teacher thought that the composition was interesting, but I had better handwriting than K. I vaguely remember that he was describing his experiences with a pet, perhaps a turtle, kept in his home. I presumed that the teacher submitted it somewhere, but I never knew the outcome. I also think K never knew what I did about his composition.

As for my own composition, I only remember one sentence. We went on a school excursion to a botanical garden in Hongo, when I was in the second or third grade. The place had a peacock in addition to various plants. So, I included in my composition after that trip a line like this:

"The peacock spread her wings as if she was showing that there was no bird so beautiful as she."

I felt that such an expression was artificial and stereotyped, but I used it anyway. The teacher or somebody praised it. That I remember only this while forgetting every other composition of mine from those days is an indication of my discomfort with doing something that didn't suit me. I never used that type of expression again.

In those days almost every pupil at or above the second-grade level brought lunch in a lunch box prepared at home. One could also order a package of sandwiches at one of the two stationery stores in front of the school gate, and pick it up at lunch time. In the second grade, the boy sitting in front of me was the son of the physician who lived next to the stationery stores. At lunch time one day, I had just opened my lunch box, when he turned around as if he wanted to examine the contents of my lunch box. He then peeled off a thin slice of cooked carrot stuck to the lid of my lunch box, and ate it. Apparently satisfied, he smiled at me, and we both started eating.

I liked his spontaneity, and so told this to my mother, who was much amused. She in turn told the story to his mother at the next parents' meeting, who was more amused than embarrassed, and, I presume, said something like, "Oh, I am so ashamed of my son's rude act, but the lunch you prepared must have looked delicious." This incident long remained in my memory as a symbol of the relaxed and peaceful mood of those school days.

Three years before I entered elementary school, there was a full-scale revision of the curriculum and textbooks of elementary school up to the sixth grade. All historians of Japanese education unanimously agree that this was great progress, and my own experiences can endorse that opinion. The textbooks of Japanese were written with special attention to the emotional aspect, which was commendable, but the editors might have overdone it, because the texts were almost saying, like the peacock, that there were no textbooks in the whole world that were so delicately written as theirs. The arithmetic textbooks and the methods of teaching were progressive, and I feel that way particularly in comparison with what I had later in middle school and even in high school, which was very old-fashioned.

The teachers were earnest and conscientious. Besides, it was only three years after this revolutionary change, and so they were very interested in absorbing the new ideas, which must have been one of the factors that made my first two years in school happy. My school life became less happy in the third grade and thereafter, but I come to that point later.

The curriculum included, of course, physical education, music, crayon drawing, manual arts, and calligraphy. Although I was short and skinny, I had no problem with sports. The same was true with music, as it was only a matter of singing songs. I hated calligraphy, because one first had to prepare liquid ink by rubbing down an ink stick, and I was very clumsy at that task. I was able to make a neat copy of K's composition because I used a pencil, not ink.

There was a problem with manual arts. I managed the job of pasting pieces of colored papers on cardboard, but there was one thing I never learned to do even passably. Namely, the teacher would give us thin bamboo sticks and cooked round beans. Then we would be asked to make some three-dimensional objects by thrusting sticks into beans. I was able to understand the idea and even able to draw a blueprint in my head. But once I started thrusting a stick into a bean according to the plan, the bean was often crushed; it never behaved as I wished. A curse on the educators who invented this wicked game!

Many years later I came to the United States and found toys consisting of various kinds of parts with which children could build many interesting objects. I often played with my daughter, and to my great enjoyment, these toys worked perfectly, and somehow I felt I took revenge for my cursed bamboo sticks and beans. May I also say, at that point I was like a rejuvenated Faust.

Sports day and performance day were the two biggest annual events in elementary school. The teachers prepared for them in the most earnest way, and the parents and families looked forward to them. The events have certainly recreational aspects, but the teachers took them to be something more than that, and expended considerable effort to make them successful. As a first grader, I was assigned a part in a drama on performance day. Two or three large cardboard boxes were placed on the stage. The part required me to wear a white costume and play a rabbit, and jump out of one of the boxes, like a jack-in-the-box. I think a girl similarly jumped out of another box. Also I had to sing a song and memorize a few lines. I used to remember the song, but now I don't. At some point I delivered a line, "I feel very honored by your compliments," but its context is completely lost.

It is my impression that such dramas in elementary school had become increasingly elaborate, and perhaps my time was the pinnacle of that trend. I participated in such events rather passively without enthusiasm; in other words I was not proud of playing a rabbit, though I think I tried my best, once the part was assigned to me. I remember that those with parts in the drama had a group photo taken with costumes on, but no parents tried to take pictures during our performance, which was normal in those days.

The reason the teacher chose me may have been that I had a clearer voice than the other pupils in my class. Some of my classmates were better than I in manual arts, gymnastics, and crayon drawing. There was a fourth-grader who was almost professional in calligraphy, with whom I was a good friend. Once in a stationery store, he explained to me the characteristics of good ink sticks and writing brushes, and was disappointed, as the store had no good ones. I believe he was better in that art than our homeroom teacher. There was also a fifth-grader who was like a superman on a horizontal bar or in a swimming pool. However, there was one thing in which I excelled: reading aloud. Our teacher in the second grade gave us a supplementary reader. One day I was asked to read a story to the class, and did so. The next time the teacher assigned the same task to somebody else, the class rebelled by claiming that I must be the reader. They liked my reading of that story so much that they wanted to hear it again. The teacher was compelled to give in and asked me to read. Then my classmates listened to my reading as if they were hypnotized. Well, strange as it may sound, that happened in the second grade.

That teacher was quite progressive, and tried various new ideas outside of the standard curriculum, such as a supplementary reader. I think he was in his early thirties at that time, and was supported both morally and materially by his friends in his attempts, though I was never certain about this point. In the next academic year he obtained a position at a women's middle school in a coastal city in Aomori Prefecture, facing the Pacific Ocean, about 350 miles north of Tokyo. He certainly improved his status, but I always wondered if he really enjoyed himself among the provincial girls of that city, who I didn't think were more sophisticated than elementary school children in Tokyo.

One day in the first grade the art teacher let some of us make free crayon drawings of whatever we liked. One of us drew a hydroplane, which impressed me greatly, as the pair of rotating propellers looked so realistic. So I said to him, "Wow, how can you draw so beautifully?" to which he replied, "Oh, I learned it in kindergarten." That made me regret my not having attended kindergarten, as I thought I lost the chance of learning many good things there. However, I am not sure whether the teacher liked such a stereotypical drawing.

I learned watercolor in the fifth grade. My art teacher would lead us pupils to back streets of the Ohkubo residential area, and would let us make plein air watercolors. Each of us would take a suitable position on the road and make his own composition. Then walking around, the teacher would make comments. I sat at the end of a long and narrow street, and made a perspective of the houses and trees on both sides. It was an Ohkubo version of a Utrillo painting. I was so foolishly honest to the point that I included a telephone pole in my composition. When the teacher came to me, he said, "That pole is obtrusive; you'd better erase it." That was great advice to me; I really felt I learned something about the stupidity of being rigid.

A large open field called Toyama-ga-hara adjoined the western border of the land once occupied by Toyama Villa, and half of it was in the range of the Kiri-ezu. It was officially intended for military drills, but was a splendid playground for us children, as we could enter freely. The Yama-no-te railroad line divided the field into two parts; the western side had thickets of trees, but the other side had only bamboo grass without trees. One day we went there and painted watercolors. On the next day, the teacher pinned each work on the blackboard and made comments, which were not just good or bad. Mine was the view, from a certain height, of the railroad and the trees on the western side, and perhaps also some clouds in the sky. No trains or human beings were in the picture. He said, "This is interesting, it looks like one by a futurist." He must have said more, but I remember only those words. I didn't know what a futurist or futurism meant; neither did he explain the word. Perhaps he appreciated that my watercolor was not sentimental and looked somewhat abstract. He was young. Probably he was seriously interested in becoming a professional artist himself. In any case, he was a free-thinking man, and quite inspiring.

Unfortunately, he was drafted, as many young men of the time were, and so the homeroom teacher, after taking a short course in teaching art, took his place. It turned out that this new art teacher was very rigid. I would do a still life in watercolor and would use a soft gray shade, which made him unhappy, or even furious, if I exaggerate a bit. He insisted that it must be in strong black. I presume he learned an idea of emphasis in the course he took, and applied it indiscriminately.

Generally speaking, however, the teaching of art is easier than that of languages or mathematics, because in the latter's case one cannot proceed to the next step unless one fully understands certain basics, whereas the matter is far more flexible in the former's case.

Here is a story about Shizuya Fujikake, who taught art history at the University of Tokyo in the 1930s. He would show various works of art in reproductions or in slides, and he would simply say, "Wonderful," "Superb," or "How beautiful," and that was all. This was told to me by the novelist Kiichi Sasaki, who took courses with him, and with whom I was friendly. Of course Fujikake must have talked a little more, but I can accept his attitude, as he showed at least what good art should be.

Returning to my elementary school, most teachers were conscientious, as I already said, but sometimes were too eager. An incident that happened when we fourth-graders went to the Yamaguchi reservoir on a school excursion comes to mind. We first went to Takada-no-baba Station, where we were to take a train to the destination. We were waiting for the train, standing next to some concrete stairs. My school had three classes in one grade. Taryo Ohbayashi, who later made a name as an anthropologist, was the class-president of a class different from mine, but we of course knew each other. His homeroom teacher, pointing to a rectangle made by one of the steps of the stairs, asked Taryo the following: "What degree is the angle made by the diagonal and the base?" Taryo answered something, but unsatisfied, the teacher said, "Is that so? You'd better think more carefully." As I was nearby, I heard these exchanges, and was quite disturbed by the fact that the teacher was not interested in making pupils relax, and only in trying to teach more on an excursion day.

Though unrelated to this episode, I recall an incident of a similar nature, which happened in Toyama-ga-hara, when I was in the fifth grade. As I already noted, there were two sides of that field divided by the Yama-no-te line. On one Sunday afternoon, I and one of my classmates were in the side that had trees. Then we saw a group of Boy Scouts wearing their standard uniforms and round hats, about eight in all, headed by an adult leader, walking towards us. They stopped at a point about twenty yards away from us. The leader first let the boys stand in a row, and instructed them to do some work, which turned out to be the following. They were divided into two groups, and the boys of one group were arranging a plan. An agreement having been reached, they began performing a pantomime. We two watched them with an intense curiosity. Making use of a puddle surrounded by some trees, they played one of Aesop's fables about Mercury, two loggers, and the axes, which was so skillful that I was greatly impressed.

When they finished the pantomime, the leader approached us, and said, "Why don't you join the Boy Scouts? That would be far better than loafing your time away, as you are doing now." He went on to describe what the good points of Boy Scouts were. I answered vaguely, but I was thinking, "I will do anything but that." I was impressed as much as, or more than, the case of the hydroplane drawing. But this time, I was in no mood for joining the regimented life at the sacrifice of the freedom of idle Sundays. Those boys looked somewhat older than us, but I am not certain. The war caused the Boy Scouts of Japan to disband about two years later. They resumed after the war, but I never saw their uniforms and round hats again anywhere. I have often wondered what happened to those smart-looking boys since then.

Though I had no interest in the Boy Scouts, I once dreamed of a better place than my school. Walking from my home in Ohkubo southward, I would find an elementary school on the left side of the road near Shinjuku. It was a three- or four-story building made of transparent glass walls, which was quite modern. Whenever I walked by it, I fancied that I would receive a better education there from modern-minded teachers. In this case I was actually seeing the building, but in the other case, the matter was completely in my imagination.

There were several elementary schools with the same poetical name in Nakano Ward which was adjacent to Shinjuku Ward. The name is *Momozono* in Japanese, which can be translated to a peach garden, but the original word has an echo suggesting something of high class. The name became known to me when I read an article that had nice words about one of those schools in a newspaper or something. That article, reinforced by the poetical sound of the name, made it an ideal elementary school in my imagination. Later I confessed this silly idea to a pupil of a different elementary school of Nakano Ward, who made fun of me by speaking contemptuously of the school. I didn't really believe in the existence of an ideal school, but I still regret that I failed to ask anybody's help for my wish of seeing the inside of that glass-covered school, and checking whether the desks were modern, for example.

As I wrote before, there were many military facilities in the Ushigome and Ohkubo districts in which my homes were located, and thus many army (and also navy) officers and generals lived nearby. Naturally their children were pupils of my elementary schools. For instance, the military school whose sports day I mentioned was the base for the military band, and the bandmaster, an army captain, was the father of one of my classmates at the Ohkubo elementary school. However, the families and children of those army or navy men were not necessarily militaristic, and for the most part not much different from those of other professions. The son of a major general was my good playmate in the first and second grades, and we two visited each other's house, and I found nothing extraordinary in him or his family.

Here is a somewhat atypical example. The son of a navy commander was my classmate in Ohkubo elementary school. His mother would wear a Western-style dress and also a showy hat when she attended school functions, which made me wonder if she had been to foreign countries with her husband. In fact, nobody cared; only, she was a little too conspicuous. A classmate of mine told me of his experience when he visited their house. He found the Japanese-style wooden veranda in the house very dirty. Normally one is supposed to take off his or her shoes before entering the veranda, but the dirtiness made him think otherwise, and so he tried to enter with his shoes on. Then the son of the navy man demanded, "Oh no, take your shoes off!" I also visited there, and found the place exactly as described.

Regardless of the professions of the families, my classmates were generally well behaved, and I had good relationships with them. Some of them went to military schools later, but my impression was that they wouldn't make good military men. I felt that way particularly about the son of the bandmaster.

Here is a small incident which shows the feeling of our mutual relationship. While in the sixth grade, I borrowed a book from a classmate. When I returned it to him, I said, "Thank you, but this book contained merely routine stuff, and did not interest me much." That was indeed so. Then another classmate who heard this admonished me later: "You shouldn't speak so bluntly; it would be more polite to say that the book was interesting." Apparently he was taught to speak that way by his parents. I said nothing, but I was not convinced. Certainly the matter depends on the situation. I have been careful particularly when I am speaking to those who are younger than I. But I think, as a sixth-grader talking to another sixth-grader, I was right in being frank. After all, I was criticizing the book, not the lender. A similar incident happened many years later, when I borrowed the memoir of the wife of one of the Canadian ambassadors to Japan describing her experiences in Japan. The lender was a good friend of mine, who happened to know her, and accepted nonchalantly my criticism that it was one of the most superficial writings of that kind. I later learned that false compliments and even false invitations expecting declination were common in Kyoto.

Returning to military men, one of my neighbors in Ohkubo was Kurazo Suzuki, a lieutenant colonel, who, as the principal official of the army's press unit, exercised tremendous authority over matters involving news information, films, popular songs, plays, comic strips, and novels. There were many writers who were harrassed and even harshly treated by this notorious individual, and so there

were quite a few stories published after the war about the wrongs
he did them.

However, as a neighbor he behaved reasonably, and nobody in
my neighborhood knew how arrogantly he exercised his authority.
He had a son who was a year younger than I. I did not like this
boy, as he was impudent and cunning. Once there was a neighbor's sports day, and it so happened that I had to deal with him.
Then he faked some "friendly" advice to me so that he could take
advantage of my naive nature. I let him do as he liked, because the
matter was minor, and I knew that I didn't have to deal with him
afterwards. Well, he deserved to be the son of that infamous father.

There was also a man who might be called an extreme opposite to military men. The poet Mitsuharu Kaneko, known for his
nihilistic and anti-war poems, lived in a house close to the Benten
shrine. Once I saw a man in untidy attire strolling along the street,
and somebody told me that the man was the poet. He had a son
who was in the same class in elementary school as my brother,
and so one day my brother and I visited his house. I found a great
quantity of toys piled up in a corner of the corridor on the second
floor, and I never saw such a scene anywhere else then or later.
The boy was an only child, and apparently the father was not the
type of man who could control himself.

In Japan today there are many obese children, but in my school
days there were only some children who were fat, but not obese.
Even the percentage of such children was small. Both Taryo Ohbayashi and the son of the poet were of that type. I remembered that
fact, because I was always conscious of my spare frame, and so such
a physical characteristic immediately attracted my attention.

There is another phenomenon common in Japan today, namely
the boys and girls who refuse to go to school. They are causing
big problems for the parents and educators. There were no such
children in my school days, though truants, if rare, still existed.
There was a truant in my class at Ushigome elementary school.
After a long absence, he showed up one day, and was able to do

three-digit multiplication perfectly. The teacher said afterwards, "Mr. X was not attending school, but did arithmetic very well, whereas you lazy boys cannot do so well even though you are getting proper instruction," which did not make much sense, of course. In fact, I never saw the boy again after that.

A classmate of mine in Ohkubo elementary school worked after school in the place where he lived. There were of course some others who helped in the family business, but this boy was not that type. Apparently he was an orphan, whose guardians did not like to be in their position, and so the matter looked like a case of illegal child labor. I don't know what happened to him after elementary school.

I insert here the caption of the photo on page 93. It was taken on a platform of Shinjuku Station on one early morning in late July, 1936. In those days many elementary and middle schools in Tokyo had summer school programs on the seashore or in the mountains. My elementary school in Ushigome used to send the fifth-graders who so wished to a Japanese-style hotel at the foot of Mt. Tateshina (altitude 8000 feet) in Nagano Prefecture. The train in the photo is about to leave for Chino Station, fifteen miles south of the hotel and one hundred miles west of Shinjuku. The parents or guardians of a pupil could stay in the same hotel as long as they wished. The faces looking out of the train windows are those of the fifth-graders, including my brother, who participated in the program. The man leaning forward at the entrance of the train is my father, who was going to spend a few days at the hotel. Those standing on the platform are the parents and relatives who are sending the pupils off. The man on the far right is the school principal, and my mother is on his left; I am the boy in front of her holding her parasol.

## 3. Our Universe

There is one thing I should note about my elementary school education in comparison with its American counterpart. It was in

1939 or 1940. I was in the fourth or fifth grade. One day, one of the teachers of the school was telling us about what constituted the universe. First of all, there is our earth circling around the sun, but it is merely one of the nine planets that similarly circle around the sun. There are also satellites such as the moon, orbiting the planets. The sun is a member of the family of innumerable numbers of stars, called the Galaxy (*Gin-ga* in Japanese, meaning "the river of silver"). He might have mentioned that there were many more galaxies outside ours, but I don't remember.

But here is what I clearly remember and must emphasize: He said that every star in the galaxy has its own planets, making a system very much like our solar system. Whether he spoke of every star or many stars is not important. He may have been vague, but in any case he told it to us as an accepted fact.

At that time there was a planetarium in Tokyo built around 1936, which was a popular attraction. Many elementary schools and middle schools had semi-official programs and most members of my class, escorted by the teacher, attended. Therefore it is possible that the above description of the universe may have been made in that context. I cannot tell where the teacher got that idea. Nor can I tell what was being taught in other schools, but I think there was nothing extraordinary in his acts. Toward the end of the nineteenth century, high-resolution astronomical telescopes had been developed, and as a consequence far more stars than in the previous centuries were discovered. This may have helped form the crude idea of the universe similar to what our teacher told us, and it is possible that most of the popular books on astronomy in the early twentieth century were written in the same spirit.

Anyway, since that day in class, I never doubted the truth of what he said. It sounded completely reasonable, and I never heard any opinion against it until sometime in the 1960s when I realized that most people in the United States thought that our solar system and humankind were unique in the universe, and there was nothing like it anywhere else. The first foreign country I visited was France, where I stayed ten months in 1957–1958. After that I went to the States and returned to Japan in 1959. I came to the States again in 1962, and I think I had that "revelation" around 1962–1963. There was much science fiction in the 1950s and 1960s in which planets outside our solar system were mentioned. But they were merely fiction which should be viewed in the same way as ghosts, vampires, werewolves, and Frankenstein's monster.

Of course in the 1960s there was no known planetary system other than ours, but to someone like me raised in Japan, the thought of the uniqueness of the solar system seemed extremely odd. Clearly it is rooted in the Judeo-Christian religious belief, and if an American teacher had mentioned the possibility of other solar systems, many parents would have complained, which may be so even today. Of course no parents of my classmates complained.

I should also note a well-known fact that the United States has been slow in introducing Darwinism in elementary school education. Even now many American citizens are in favor of teaching so-called intelligent design, and the President of the United States shamelessly supports the idea.

There is one more strange aspect of the thinking of American scientists. Namely, the belief that only human beings have the faculty of thinking, and no other living things do. If a scientist shows an example contrary to that idea, the believers always say, "That cannot be called thinking," by making very narrow criteria. That has been changing in the last ten years or so, but there has been no such rigidity in Japan. I think this is also caused by the same kind of religious belief. I feel very happy that I was never under the influence of such religious dogmas.

Although this is unrelated to religion, let me note another thing which I took for granted in Japan, but which was not so in the United States: a swimming pool in an elementary school. Such was common and almost standard in Japan at least in the late 1930s, and I always thought that the same was true in the States

or in Europe. But apparently it is not so in the States even now.
I never checked this point in the case of Europe.

Overall, I have few complaints about the education I received in
public schools through the tenth grade. Not that I liked everything
about my school life. There were several things I didn't like. One
is related to national holidays. There were two types: one was just
an off-day like Sunday for everybody. The other was half ceremonial. A ceremony commemorating something important, which all
pupils and teachers attended, would be held in the morning. After
that it would become an off-day. I don't remember the exact program of the ceremony. It certainly included the singing of the
national anthem and a speech by the school principal. More importantly, there was something which we might consider bizarre or
silly nowadays, but which we were supposed to take very seriously
at that time.

To explain the nature of this rite, it must be noted that each
school had a photo of Emperor Hirohito given by the government.
(It may have been a double portrait of Emperor and Empress, but
that point is not important.) It was a sacred object of worship
and was stored in a fire-proof structure called *hoh-an-den* which
looked like a small Shinto shrine. On every ceremonial holiday the
principal would take out the photo from that shrine, and place
it in an alcove covered by a curtain in the hall where the ceremony was going to be held. When the time came, the principal
pulled the curtain, exposing the photo. Then, at the word of com-
mand given ordinarily by the assistant principal every attendant
made a deep bow to the photo. The principal then read "The
Imperial Inscript on Education" (*Kyo-iku Choku-go* in Japanese)
decreed by Emperor Meiji on October 30, 1890, which took three
to five minutes, depending on how fast or slow it was read. We
remembered the date, because it was read at the end of the document just before the very last words "the Imperial Sign and Seal."
Either after or before this we would sing "Kimi-ga-yo," the national anthem. The principal would make a speech about why

that day was significant. We would sing another song made especially for that particular holiday. The program of the ceremony was roughly as I described, but my ordering may not be accurate. Eventually the principal would close the curtain.

After the ceremony in the hall, we would go to the classrooms, where the homeroom teacher would make another short, supposedly enlightening, speech. At the end each of us would be given a pair of dry confections made of sugar, one pink, the other white, in the shape of 16-petal chrysanthemum flowers, which was, and still is, the imperial crest.

The poem on which "Kimi-ga-yo" is based used to be sung in the twelfth to fourteenth centuries by a professional female dancer to entertain a person of high social status, who in turn would give her a gratuity. I find it extremely odd to make it the national anthem. Maybe we school children received dry confections as gratuities.

The whole ritual was boring and nonsensical. Of course one can justify such ceremonial holidays, but I don't see any point in such formalized bowing and reading. Nobody liked it, and everybody knew that we were forced to behave that way. I think most, if not all, of the principals felt the same way, and they made sighs of relief at the end of the ceremony. Indeed, anyone who failed to do the job properly might have been fired or demoted. It is the same story as in a communist country like North Korea. Besides, I am sure Hirohito knew that this was done on every holiday and in every school, and he never uttered any word of regret, which disturbs me greatly.

There was an incident of a similar nature that happened in January 1948. Every new president of the Upper House and new speaker of the Lower House of Japan would present himself to Hirohito. When the official left, it was the rule that he must keep facing Hirohito, so that he had to walk sideways like a crab. One official refused to do so. This is called the affair of the side crawl of a crab. Even then, Hirohito and his subordinates were happy to enforce such nonsensical court etiquette on those important persons even two years after the war. There were many other types of forced bowing to him, one of which is bowing from afar, so silly that I would not describe it here. It would make a book of fair length to discuss all such forms of worship practiced in those days.

This may be no place for discussing his war responsibilities, but let me note an important fact that is almost forgotten today. Besides being the emperor, Hirohito was His Majesty the Generalissimo, to whom all military personnel reported. This was completely outside domestic affairs, which required consultation with the ministers.

There was a card game played by children in those days consisting of two sets of cards each of which was marked by an army symbol with ranking. The card of the general beats that of the colonel, but is beaten by the battle flag, which in turn is beaten by the sublieutenant, who holds the flag. There was no card for the generalissimo, as such a card would have made the game meaningless. Hirohito was in such a position, and so it is impossible to say that he was not responsible. This fact has already been pointed out by a scholar of war history, but I think it is worth mentioning again.

On a lighter note, I must add one aspect which cannot be found in a communist country, but could happen in Japan. Though the whole ritual was done very seriously and nobody made fun of it publicly, many of us instinctively perceived the hypocritical pretense in these theatrical rituals. In fact, we children played a game with that sacred "Imperial Inscript." This document begins with a proclamation "I think, our Imperial ancestors (pronounced "kohso-kohsoh" in Japanese) established our nation in the remotest days, and you loyal subjects, blah, blah, blah." So we would play a game as follows. It is played by two children, say, players A and B. First A, pretending to be the Emperor, reads the script, and orders B to imitate what he does. He begins with "I" putting a finger on his nose. B does the same. Then A says "think," with his arms crossed. B does the same. Then A continues "our" (without

Imperial ancestors), stretching his both arms widely. B does the same. At that moment, A tickles B at the armpits, saying "kohso-kohsoh." The point is that "kosoguru" is the Japanese verb for tickle. Anyway there would be laughter, and of course the game cannot be played twice on the same person.

Such was no secret. Those principals must have played the same when they were schoolchildren. No undercover agent was watching us.

There were also vulgar versions of the Imperial Inscript. One of them goes like this: "In our opinion, when we break wind, you subjects are overwhelmed by stinking; holding our nose, Imperial Sign and Seal."

This is not surprising. In any country at any time such is common. It is often said that a great majority of the Japanese population worshipped Hirohito as a living god, but that is completely wrong. It is true that the majority behaved *as if* he was that kind of being, but one cannot go much beyond that.

There were several jokes spread among the citizens, even in the darkest days in modern Japanese history. As mentioned earlier, on February 26, 1936, a group of young army officers attempted a coup. Though it failed, they managed to assassinate two ministers and a high army official. It is said, when Hirohito heard this news from a court official, he staggered. Naturally the official asked, "Oh, Your Majesty, what is the matter with you?" He replied, "I lost Jushin." "Jushin" is a Japanese word, taken as "important ministers" and also as "the center of gravity." It may be added that an ad-balloon was sent up to urge the rebel army to surrender, which was no joke.

There was another thing which I did not like in my school life: in the winter, the schoolroom was not well heated. I am certain that American schools were better heated at that time. I already mentioned the chilliness in most Japanese homes. In fact, how to deal with physical coldness has been an important problem in my life, but I come to this point later.

Some educational movies were shown in school occasionally. The screen was hung in front of the alcove where Hirohito's photo was placed. I remember only those about hygiene. One is a story about a boy of age seven or eight who buys cheap sweets that are filthy. There is a close-up scene of many big flies walking on the confections. Eventually he contracts dysentery and dies. The other concerns an older boy who works, for what reason I don't remember. He carries heavy objects in a wagon in the rain. He contracts pneumonia and dies.

That may be an oversimplification, but in any case these were depressing, and not much better than the photo of Hirohito. However, it must be noted that pneumonia and various kinds of contagious diseases were certainly principal killers of children in those days, and therefore the movies had some good points. Later, one of my good friends, a classmate in middle and high schools, told me that when he attended an American elementary school, he was always cautioned about pneumonia, which people viewed as a main cause of death.

## 4. Weights on a Child's Mind

I was very skinny as I already noted, which caused my parents anxiety. I was near-sighted too. The elementary schools in those days had a program of letting those pupils who wished take cod-liver oil drops. So my parents put me in the program. It was like the vitamin pill of today, and I guess it was good for my health in general terms, but of course it didn't add any more weight to my body. However, I had no problems with gymnastics, as I said before, and I was able to swim too. Therefore that I was skinny didn't bother me.

Still, many years later I had a bitter experience because of my skinniness. It happened when I participated in a four-week conference held in Woods Hole, Massachusetts, in the summer of 1964. There was a public bathing place where the participants of the conference and their family members could swim. A floating platform was placed about 80 yards away from the shore, and my wife told me proudly that she had reached there and swam back. I said, "Oh, that's nothing," and jumped into the sea water. In fact, several years earlier, I had experiences of swimming far off-shore the Izu Peninsula in Japan, and of lying sprawled on the sea surface. But, it was my unpleasant surprise that the water of the Atlantic Ocean was very cold. Anyway I reached the platform and took a rest there, but was trembling in the wind. It made no sense to stay there, so I braved the task of swimming back. Although I was able to reach the shore, I thought that I would be drowned. Luckily I did not contract pneumonia, but at that time I determined not to swim anymore.

The matter would have been different if I were not so skinny. My skinniness caused a problem of a different kind when I was a child. Before narrating the story, let me note the desire I had in my childhood: to become an adult as quickly as possible. I don't know when I began to feel that way, but I assure the reader that I had no attachment to my being a child. After all, if I were an adult, I wouldn't be forced to thrust thin bamboo sticks into beans! I cannot recall exactly how I felt; possibly I thought the adults were manipulating the children to their advantage, but that is of course an exaggeration. It is certain, however, that in my opinion at that time I was losing greatly by not being an adult.

There was an occasion when I felt they were exploiting me. My eldest sister graduated from a girl's school, having finished the eleventh grade. She was taking a course in dressmaking some-where. One of her first projects was a sleeveless shirt on which some colorful designs were stitched. My mother tried to make me wear it to school. It was some time in June, and the weather was not yet so warm. I said, "No," because no classmate of mine wore such a showy shirt, and besides, more importantly, I would feel chilly for the reason I already explained. But they did not relent; they said this was a splendid shirt, and you would look wonderful in it, and so on. They invented every reason why I must wear it, and finally succeeded in coaxing me to do as they wished. I really felt unfairly treated by them. Maybe I should have faked a cold or pneumonia, but I didn't have the courage.

In those days, summer vacation was from July 25 through August 31. When I encountered summer vacation for the first time in the first grade, I was very puzzled. My father still goes to work, as other fathers do. But pupils attending school suddenly stop doing so; the same is true with my sisters and brother. Why? Not that I liked school so much that I wanted to continue going. Simply, I did not understand the reason for summer vacation. However, I kept this question to myself, and never asked anybody about it. If I had asked, the answer might have been, "You are a disagreeable boy who asks such a strange question unsuitable for children," and perhaps I instinctively knew that.

It must be pointed out, however, that it is not easy to answer this question. There are historical and practical reasons for summer vacation. It is true that learning capacity decreases in warm weather, but that is not the main reason, as summer vacation exists even in cool countries.

But many years later I questioned the meaning of summer vacation for a different reason, when I was a thesis advisor of graduate students at Princeton. Many of them had no experience in research, and did not know that a long vacation with no work would erase their memory very easily. Therefore before summer vacation I would tell them how to deal with this danger, and also that what I was telling them was extremely important. That made me feel that I was an aged man, and that my job of advising would be much easier without summer vacation.

There is another matter that troubled me when I was a child. In elementary schools many teachers said that we must appreciate the hard work of farmers, who produce many kinds of things we eat every day, such as rice and vegetables, and I accepted it as an important fact. They also said that we should not waste goods, and should be thrifty, which sounded very reasonable. One more teaching: "We should not put much weight on the importance of money in our lives." I also accepted this too, but in reality the children did not know how important money was, and so what they said made sense only in a special context, or was hypocritical.

However, there was an interesting aspect in my reaction to what my teachers said about the work of farmers. My teachers were saying in essence, "Don't forget the existence of the laboring class." Although this is not completely accurate, I was influenced by this thought to a considerable degree. For example, my parents and I would go on a pleasure trip by taking a train; then I would see from a window of the train some farmers indeed working in a field. The sight would make me feel guilty of enjoying leisure time without working.

My family was not very affluent, but certainly did not belong to the laboring class, and that fact bothered me. Our destination would be an amusement park, a famous Buddhist temple, a Shinto shrine, or a scenic place. Once there, I would lose the feeling of guilt, and would enjoy myself as an ordinary child. Still, I was unable to flee completely from that feeling during my childhood. However, I never uttered a word about it, keeping the thought to myself. Somehow I was ashamed of telling such an idea to anybody. Also I did not like to look hypocritical, though the word hypocrisy may not have been in my vocabulary then.

While in middle school during the last period of the war, we were forced to work in a factory that made parts for fighter planes, and at that point I knew the meaning of the labor in such a place. My sense of guilt completely disappeared by the end of the war. Perhaps my experiences of various kinds at that time and later periods helped me lose that sense, but rather, that was inevitable because at that time I belonged to the most straitened class, if not the laboring class.

When our life became stable after the war and I was teaching, I told this sense of guilt in my childhood to my colleagues and other friends of my generation, and inquired whether anybody had the same feeling. I was expecting at least some affirmative answers, but to my surprise, all the answers were negative. But, we all agreed on one point: even though we did not belong to the proletariat, we belonged to the exploited class.

I now think that it was absolutely unnecessary for me to have feelings of guilt. It was someone else who should have felt that way, but I realized it only after the war.

# CHAPTER II

# AS A STUDENT

## 5. At Middle School

In contrast to my happy first two years in elementary school, the
next two years were not so enjoyable. The new homeroom teacher
was not terrible, but to me he looked like a man lacking in sincerity,
but going along cleverly in the world. I did not like him anyway.
In my mind, every school day in that period was rainy and the
classroom was gloomy, though of course that was not necessarily
so. As for the fifth and sixth grades, I was in Ohkubo elementary
school, and the homeroom teacher was the person who insisted on
a very strong shadow in watercolor, but he was tolerable otherwise.
I was not unhappy, as I got along well with my classmates.

By 1940 the Japanese army was bogged down in China making
no progress. Every ordinary Japanese citizen felt the shortage of
everyday goods, but strangely the army did not care. In 1939
there was a military conflict at the border of Soviet Russia and
Manchuria, and Japanese troops suffered heavy damage, but the
details of the whole affair were kept from the Japanese public.
However, a few years after this defeat, an army captain named
Sakae Kusaba wrote a book that recorded the misery of the battles,
from which the reader could infer how disastrous the matter was.
I read the book when it was published, but did not realize how
bad the situation was. But nobody was able to stop the fatal force
which eventually led the country into war with the United States

G. Shimura, *The Map of My Life,* doi: 10.1007/978-0-387-79715-1_2,
© Springer Science+Business Media, LLC 2008

and the British empire on December 8, 1941. It is not my aim to
write about the war, but I mention these facts for the purpose of
showing a historical perspective of the time.

I should mention, however, at least how I reacted to the news
of that fateful beginning of the new war. Strangely I don't remember much about the Pearl Harbor attack, but I clearly remember
the sinking of the British battleships, the Prince of Wales and
the Repulse. That was thoroughly thrilling. I thought, "They at
last learned a lesson," and I never regretted having had such a
thought. At that time the British empire occupied an incredibly
large area on the world map, and they had been engaging themselves in whatever suitable, such as the Opium War, for their evil
colonial policy. It was as if the opening sentence of *The Decline
and Fall of the British Empire* had been written.

Anyway my six years in elementary school ended in March,
1942. At that time I felt relieved and thought that those were
very long six years. In those days, there was only six years of compulsory education; there were also vocational schools in addition
to middle school, and therefore relatively few sixth-graders went
to middle school. There was a tatami maker near my home, whose
son was the top student in the class different from mine. After our
graduation from elementary school, I would observe him making
tatami in his father's shop, which caused me to have uneasy feelings, as he was so good in school. It is possible, however, that he
attended an evening class somewhere. Be that as it may, in the
years before 1942, the middle school entrance examination was a
trying barrier which those sixth-graders must get over.

There were preparatory schools and private tutoring; guidance
in this matter was a nontrivial and time-consuming job for the
teacher. But fortunately the whole examination process was sim-
plified the year I entered. The report on the pupil's school record
and an oral examination were the only requirements; there was
no written examination. The school of my choice was called The
Fourth Tokyo Prefectural Middle School, and would have been in

the range of the Kiri-ezu, had the map been one inch wider. There
were 300 applicants, of which 250 were accepted, and therefore the
ratio of pass to failure was unusually high. As to the oral examination in my case, I remember only: "How old do you think you
will be when you become independent of your parents? How old
will your father be then?" My parents were afraid that I might be
rejected because of my skinniness that made me look unhealthy,
but I was accepted after all.

Actually, getting in was the easiest part of the matter. After a
week or so I began to think that I was in the wrong place, as the
school was very rigid and strict, to an almost suffocating degree.
From the viewpoint of the school, they were eager to train the
students to conform to their ideal type. In principle, the same was
the case in every middle school. For instance, the middle school at
the end of the slope I mentioned before might have accepted boys
who were not smart, but once they were in, I think the school
was not very lenient, and in fact strict in their own way, which
may have been what the parents expected. But "the ideal type"
depended on each school. Some were certainly modern and liberal,
whereas my school was very old-fashioned and conservative.

In spite of the rigidity, I must say that the teachers were earnest
and driven by professional conscience, so that their teaching was
not perfunctory. Also the first two years and eight months at
middle school was the only period of my school life in which I was
a docile and diligent student. That period was cut short by the
student labor mobilization, which I narrate later.

Here is how various subjects were taught in my middle school.
In horticulture, we were taught about the varieties of fertilizer, and
worked in a small field with an area of an acre, which was a few
minutes walk from the school. I found this useful. As for music, it
should first be noted that the level of Western music in Japan in
those days was unbelievably low. Returning to elementary school,
in the sixth grade we learned chords, and sang two-part cho-
ruses. The young female teacher at Ohkubo school taught us with

enthusiasm, but sadly our performances were always far below her expectations.

The textbook of Japanese in the fifth or sixth grade included the famous story about Beethoven's Moonlight Sonata. The elementary school I attended had an electric record player, which was connected to speakers in most classrooms. So, one day we listened to the sonata in our classroom. At the end, the teacher, the one who had the wrong idea about watercolor, said, "Why is this incomprehensible stream of sounds good music?" We pupils were all silent.

Of course the school was not the only owner of such a record, as there were several record companies selling records, often in sets, of the standard repertories of Western music. There were also families with organs or pianos, and some children learned how to play them. However, there was no good environment for encouraging children in music, and so in most cases, the playing did not go much further beyond that of "Chopsticks." It must be noted, however, that whether one is able to play the piano is irrelevant to one's understanding of music, as there were several famous composers, including Richard Wagner, who lacked that ability.

The music teacher in my middle school was a good musician who was well known at that time. He told us about sonata form, and explained the structure of a music piece, taking as an example Beethoven's Pastoral Symphony, played on a record player, which made great sense to me. But one day, my classmates were bored in his music class, and made some noises. He became upset and left the classroom. Then the class president, after begging the students to behave, went to apologize to the teacher, who eventually came back and continued to teach.

Having three sisters, I knew to some extent what was being taught in girls' schools, where, I think, no teacher would have been upset in that fashion. There were few problems in other disciplines, though on warm afternoons, the students naturally became sleepy, which sometimes caused the wrath of the teacher and even an unpleasant outcome, a common phenomenon in any school in any temperate country.

From the eighth grade on, we always had a supplementary reader in English in addition to the regular reader. In the eighth grade our supplementary reader was *Aesop's Fables.* In those days there was something called *anchoko,* which struggling students used at home as a cheap kind of study aid, and which, the teachers warned us, would do more harm than good to the students. However, I somehow acquired a book that looked like anchoko for *Aesop's Fables* for reasons I don't remember. It turned out that I did the right thing, as I enjoyed reading the book, which contained more stories than our textbook, and I certainly learned a great deal. Possibly, it was not a typical anchoko.

One of the English teachers was quite a character. He was in his late fifties, known in a certain literary circle, and had published a book with a good publishing company. He told us that we should forget about Germany; instead, we should pay more attention to England. It was in 1944, when Japan had the triple alliance with Germany and Italy, and was fighting England and the United States. That is what I remember; he didn't mention the United States.

We had *The Arabian Nights* as the supplementary reader with this teacher. One day I had reason to see him, and I visited him in the teachers' room, where almost all teachers had their desks. When the matter was settled, he brought out a thick volume, which was a version of *The Arabian Nights,* not really *The Thousand and One Nights.* He showed me its frontispiece, which depicted in color the slave girl who served Ali Baba, belly-dancing with bare abdomen. Then he grinned at me, as if he was asking, "What do you think of this?" The picture was ordinary and had no suggestive element. I said nothing and showed no emotion, which might have disappointed him. In any case, he closed the book, and I left. Maybe I should have said, imitating Professor Fujikake, "Wonderful" or, "How beautiful."

Once we had dictation with another English teacher. The text, about fifteen lines in all, described something about maple syrup. The teacher collected what we had done, but for some reason unknown to us, directed a student to return the papers, as he would not grade them. I was almost the last one who was leaving the room. Before leaving, I looked at the top sheet of the pile of the collected papers, and was surprised by its perfectness. As nobody was watching me, I impulsively jotted down on the top line, "Very good indeed!" I knew that the teacher used that phrase. I had no intention of making fun of the author; nor did I tell this to anybody. Of course I was unhappy with the fact that such good work was not acknowledged, but may I say, I belong to the type of people who do such things impulsively for no reason from time to time.

Over all, the level of education I received in my middle school was not low, as evidenced by a story one of my classmates later told me. During the war many residents of Tokyo fled to the countryside in order to avoid the American air raids. His family did so, and he was naturally enrolled in the middle school of the city where they relocated. He found the level of curriculum far below that of our school in Tokyo; besides, the students were uncivilized.

Classes in mathematics were not very interesting. We again learned arithmetical operations of fractions and decimals, which was all right. But we were asked to solve artificial arithmetical problems without using algebra. One typical problem is called the *tsuru-kame* problem. *Tsuru* is the Japanese word for crane, and *kame* for tortoise. Given the total number of cranes and tortoises and the total number of their legs, one is asked to determine the number of cranes and that of tortoises. The point is that one must solve the problem without using algebra, that is, without introducing indeterminates $x$ or $y$. There were many more complicated problems. In fact, such problems had been included in the entrance examination for middle school in previous years, and therefore those who took such an examination were forced to do them twice.

I never found such problems interesting. In fact, the editors of the new revolutionary textbooks in elementary school I mentioned before knew the futility of teaching that kind of arithmetic. But that idea did not reach the middle school teachers.

This kind of conservatism was not confined to middle school. I would later discover, to my disappointment, the same tendency in high schools and universities. The stupidity of the tripos in mathematics in Cambridge has been much discussed, and I am inclined to think the Japanese system had been influenced to a great extent by the antiquated British system.

But it is very difficult to throw away what is supposed to be taught. The subject must be taught only because it has been taught for a long time; the teachers never ask the fundamental question of why it must be taught. Plane trigonometry and spherical trigonometry were in the standard middle school curriculum in those days. Spherical trigonometry may be taught in some special schools, but is unnecessary as a subject in ordinary middle or high schools. However, there is an easy and basic theorem in spherical trigonometry that requires no technicalities. Namely, the fact that the sum of the interior angles of a spherical triangle minus twice pi is proportional to the area of the triangle. This is one of the easiest theorems of non-Euclidean geometry. Therefore, to teach this in the tenth or eleventh grade is a sensible thing to do, I think.

Returning to my first year in middle school, each of us was required to own a slide rule, which was beneficial. Indeed, mathematics is not just formal logic; it requires an intuitive sense of what each mathematical object is. This can be said at advanced levels, but also at very elementary levels, too. I think the slide rule helped the students gain a good sense of decimals and approximation, though probably it is unnecessary to reinstate it in today's middle school education.

Why did they insist on teaching such an outdated arithmetic? I cannot speak for them; I can only guess their thinking. Probably they rigidly believed in the hierarchy of mathematical subjects, and thought that one had to learn a subject completely before proceeding to the next higher subject. Namely, arithmetic before algebra, Euclidean geometry before analytical geometry, which must be finished before calculus. Indeed, they spent an incredible number of hours in teaching conic sections.

I just mentioned Euclidean geometry, which was being taught persistently in pre-war middle schools in Japan. Those who developed the curriculums of middle and high schools never asked the question of whether it was really important.

I remember an odd incident in middle school, which is unrelated to the question of what to teach. There is a book titled *Mathematics for the Millions*, written by Lancelot Thomas Hogben. It was published in 1937 and very popular at that time. After a few years, it was translated into Japanese. Then a certain Takeuchi wrote a book in Japanese whose title was exactly the same as the title of the Japanese translation of Hogben's book. This was already bizarre and even unethical, but besides, at one point he claimed to have found a chemical procedure of transforming mercury into gold. This was reported in the newspaper. He was a professor at a school in Tokyo whose graduates were mostly teachers of chemistry, physics, and mathematics at middle schools.

Shortly after the newspaper report, we students were told that this professor would talk to us on some topic, and indeed one day he gave a lecture, which was quite mediocre, and worse, had no meaningful content. I can ignore that aspect, but at that time I thought it very strange that the school allowed an alchemist to give a lecture. I say this because the lecture was given after the publication of his book and the announcement of his great chemical achievement. But apparently our teachers didn't care, and introduced him to us as a celebrated scholar. Of course this kind of thing can happen in any country at any time, but that it occurred in my school was shocking.

I read both Hogben's book and Takeuchi's, and remember nothing about the latter. As for the former, I was bored. Perhaps

I should have given up on it after the first thirty pages, but didn't. I continued reading with the expectation that eventually I would find something more interesting in later chapters, but that never appeared. After all, I became a mathematician, and the book was not intended for such a person, and so I should not complain.

I had some happy experiences in my years in middle school, as most boys of that age normally would have had. But we were always under the thick psychological clouds created by the war, which made it impossible for us always to be in a sunny mood. Military training was in our curriculum, and we were unable to enjoy it, or dodge it, of course. The war influenced our education in a more essential way, as there was the so-called student labor mobilization, which required us to work at various places for the purpose of helping the war effort. The students were not the only people who were labor mobilized. There was a law that all male adults and unmarried women of certain ages must be employed in a similar way. For this reason my second eldest sister had a desk job at a military establishment.

Speaking about middle school students, we did not attend school from some time in November, 1944 until the end of the war. As I wrote earlier, we first worked in a factory that made parts for fighter planes. To avoid American air raids, that factory was relocated to the countryside, and so we went to work in other places, but I do not go into details of everything we did, except to say that I acquired various kinds of odd skills that I would never have learned, had there been no labor mobilization. So far I avoided the description of the darker aspect of my life in this period, but in the next section I have to face it honestly.

## 6. The Days Before and After the War's End

My years in middle school roughly coincide with the period when Japan was fighting the United States. The war with China had started on July 7, just before summer vacation in 1937, when I was a second grader. The war ended on August 15, 1945. On that day, we were liberated from the student labor mobilization, and we resumed attending middle school not long after. The school building had been burned down, and so we temporarily used the nearby elementary school building that had escaped fire.

Before narrating the American air raids I experienced, I first mention two unforgettable incidents in the months just after the war. My parents and I were living in an apartment in Mitaka, a town eight miles west of Shinjuku. I was commuting to middle school via the Chu-oh line, but had to transfer from the local train to the express at Shinjuku Station. One day, perhaps in early October, I, together with a classmate, was standing on the platform waiting for the express. We were wearing our school uniforms and school caps; in addition a piece of cloth was sewn to the chest of the uniform. The piece showed the name of the student and the homeroom number. Those were school regulations during the war, and I believe they were not required after the war. However, we had nothing else to wear, and so we were doing as we had been doing before.

I suddenly noticed a boy about our age just a few yards away. He was wearing exactly the same school uniform and cap; he even had a piece of cloth of the same type with the same homeroom number. My companion also noticed him, but we were unable to recognize him. No such student existed in our class. The boy noticed us too, and all three stared at each other for a few seconds. Then he turned around and fled; he hid himself in the crowds, and we never saw him again.

Clearly he was a fake student. My middle school was one of the best in Tokyo, but it was bizarre to find a fake tenth-grade student. What was his motive? I was never able to find a convincing answer to that question. Of course it was almost impossible for someone to act that way during the war.

Some time around 1953 I noticed a girl in the classroom at the University of Tokyo, where I was teaching. After my class she came to me and asked a few questions. I saw her a few more times later; I was also told by my colleagues that she was attending their courses. At that time female students were relatively few, and so her existence was easily known to many people. Her main objective may have been to find a highly desirable husband, which she almost achieved, as I learned much later from an article in a weekly magazine. According to the article, a professor, in the capacity of a matchmaker, checked her background, but found no record in university files. Soon those involved in the matter or who had met her knew that she had been faking for more than four years. The matter seemed sadder, when some professors vouched for her competence in class.

Another incident occurred around the same time as the fake middle school student. One day, perhaps Saturday or Sunday, I was reading a book or a newspaper in the apartment room, when something outside boomed, and the sound repeated without stopping. Then I heard a few people talking. I went out, and immediately found out what was causing the sound. There was an electric pole at a crossroads near the apartment. I saw a man standing on a wooden crosspiece in a high position on the pole, and blue smoke was rising from his head. He was the superintendent of the apartment. The people on the street watched the man being electrocuted, but nobody shouted, or was agitated; they were just quietly talking without showing much emotion.

At that time power failures were common. Apparently he had some knowledge of electricity, and he thought he would be able to fix the problem. I think he had done so on a few earlier occasions, but on that day he failed. It was a tragedy, of course, but people, including myself, accepted it as if it were a common occurrence. To understand why people were so impassive, it must be remembered that during the war we were living next to death, and it was common to see corpses.

There were many air raids, and I clearly remember three major ones: March 10, April 13, and May 25, all in 1945. The last two burned down the houses in which I lived. As for that of March

10, though it spared my Ohkubo house, it left the largest casualties of all such air raids, except for those caused by the atomic
bombs. At that time my brother, a labor-mobilized student, was
working somewhere in downtown Tokyo, the hardest hit area, and
my family was anxious about him. So the next day, I went to his
workplace alone.

Fortunately, the train was running, and I was able to reach
Kanda Station, and from the platform there I gazed at the eastern
side of the station, but there was nothing to be seen but an enormous flat field. I walked about four miles from there to the other
side of the Sumida River, where I was able to find my brother. I
remember what the place looked like, and also that I chatted with
his classmates, but strangely, I have no detailed recollection of the
streets I walked. I vaguely remember seeing several corpses, but
what I saw a few days later left a stronger impression on me.

I was walking on Ohkubo Dori towards the place where I was
working at that time. The north side of the avenue was the wall
of the military school whose sports day I mentioned before. There
were a few corpses of those burned to death, left on the road close
to the wall, and I walked by them three or four more times. Some
thirteen years later I visited the ruins of Pompeii, and went to the
museum there as every tourist did. The posture of plaster casts
of the victims immediately made me recall what I had seen on
Ohkubo Dori.

Skipping the description of the air raid of April 13 that burned
down our Ohkubo house, I will now explain what happened on May
25. It was not difficult to extinguish a single incendiary bomb, provided one covered it quickly with enough sand and soil, and I did
so for a few bombs. However, hundreds of them were falling down,
and they started burning under the floor after piercing the roof.
My family and neighbors soon realized the futility of fighting the
bombs, and decided to flee. The American bombers had already
left, and no more bombs were falling. Our problem was how to
reach a safe place by avoiding the fires that surrounded us. We

somehow managed to find refuge in the grounds of the army hospital I mentioned earlier, though I don't remember which route we took.

The American bombers at that time were B29s. The number of B29s used on May 25 is 502. There was also a raid on May 24, which had no effect on me or my house; this one used 562 B29s. Each B29 carried more than six metric tons of ordinary or incendiary bombs. The casualties of each raid were also recorded. The largest number of dead, about eighty-four thousand, was caused by the raid on March 10. The next largest number is that on May 25, which is officially 3651. I just mentioned the major ones, but there were many more raids not only on Tokyo, but also on other cities. These numbers tell the reader the extent of the air raids. I always wondered about their strategic significance, particularly about that on March 10. Wasn't it intentionally designed for the purpose of killing as many civilians as possible?

The first American air raid was carried out by several carrier planes on April 18, 1942, just two weeks after my entering middle school. It was Saturday, with the afternoon off, and I was walking from school to my house around noon, when the planes came. A few houses one block away from my home were burned down. There were no more air raids until the fall of 1944.

Although unrelated to the air raids I experienced, I insert here what I think about the atomic bombs dropped on Hiroshima and Nagasaki, more precisely, about what Americans always say, "They can be justified because they were necessary." All American historians treated this matter as if they were a single event, without separating one from the other. I think this is completely wrong; it is absolutely necessary to discuss each bombing on its own, separating it from the other.

They dropped the second one on Nagasaki even after seeing the devastating effect of the first one. That is beyond my comprehension, and there is no justification for the second bomb. Therefore it is highly plausible that there was a hidden motive for the second bomb, that they would not disclose. We can then infer that there was the same hidden motive for the first one. It is my conclusion that the one dropped on Nagasaki cannot be justified. I would say the same for the first bomb, by inference. I don't know whether somebody already expressed the same opinion. If it were so, I never heard any opinion counter to it.

Returning to air raids, I must note that women's clothing was much affected by them. Of course kimono was impractical, but few women had trousers. There was *mompe,* which looked like pantaloons, and used to be worn by rural women working in a field. So most women wore mompe combining it with the upper part of a kimono and cutting off the lower part. Also in a certain period most of us slept with our clothes on, so that we could fight the bombing or flee without losing time.

Air-raid shelters were everywhere. We dug one at our house. Every time we moved to another house, we dug another shelter. We were all shelter-diggers. In the earlier period, whenever we heard a siren warning us of approaching American bombers, we ran into the shelters, and got out of them when the siren sounded the all clear. But after realizing that we had to deal with incendiary bombs, we did not get into the shelters; instead we stored the essential necessities there in advance, and added some more at the siren, closed it with a lid, and covered the lid with sand and soil. Then we fled, carrying a shovel. Afterwards we would sleep in the shelters like cavemen.

In the aftermath of the air raid of May 25, I witnessed a weird scene that I would never see again. The trees in the bombed area naturally lost all their leaves and small branches, but the burned trunks and larger limbs remained. That much was not strange. Within a few days, however, all the blackened trunks and limbs were covered with bright orange mushrooms, which looked poisonous, and which nobody could take as a good omen. Thus we lived in the shelters surrounded by such unearthly trees.

We remembered most of the stored objects, but a few unexpected objects managed to survive, as we were always in a hurry. I think it was in the summer of 1981, long after the war. I was searching for something in a closet in my mother's house, where I once lived for several years. I was unable to find what I wanted, but found a wooden box which I did not recognize. Out of curiosity I opened it. The box contained a set of porcelain pieces: five tea bowls, a teapot, and a water cooler. I did not remember seeing them before, and showed the set to my mother, who said, "Ah, this is one of my wedding presents." They were Kiyomizu ware decorated with a plum tree design in underglaze blue and a bit of overglaze gold and red, in a refined antique style that was not often seen on present-day pieces. As I riveted my eyes on them, she said, "Carry them away, if you like." She was eighty-three at that time; my father had died nine years earlier. I cannot tell whether she had some attachment to them; perhaps they were stored simply because they were not for everyday use, or both. In any case, I think I used them more times than she ever did.

The air raids made everybody submit to a whim of fate. Some fled to the countryside, but were bombed at their new locations. Many of the elementary school children in Tokyo were relocated. But at some point some of them reached the age at which they were no longer "pupils," and so returned to Tokyo, where they suffered the same fate as those who stayed. There were innumerable stories of the houses which survived the fire that burned down those in front of theirs, told by both the lucky and the unlucky.

The two elementary schools I attended were both burned down. There was an elementary school at the edge of Toyama-ga-hara, which was also burned, but the small structure housing changing rooms next to the swimming pool was left intact. So a few teachers lived there even after the war. Many such situations were inevitably created, as the difficulties of finding dwellings persisted long after the war. There were those who lived in the offices at the schools where they taught, even in 1949. There was a large Shinto shrine in the western suburbs of Tokyo that had a few subordinate shrines in the precincts. One evening around the same time, I saw a few people eating in one of those small shrines under an electric light, and I said to myself, "Oh, gods are behaving like that," and I wondered if I was dreaming.

I now narrate other aspects of my life in the four months before the end of the war. My family made light of the situation, or rather, was unable to find a good place to relocate, and enjoyed unpleasant consequences. I was a labor-mobilized student, and worked on odd jobs. In the short period immediately after the air raid of April 13, I was helping repair overhead wires of streetcar line No. 13. Thus I was still confined within the domain of the Kiri-ezu.

One day, our team consisting of two professional repairmen and four students, was working near Nishimuki Tenjin, the Shinto shrine mentioned in the first section. After lunch, we students were taking a rest within the precincts of the shrine; the repairmen had gone somewhere for their lunch. Growing tired of waiting for their return, we decided to leave. Two of us went in the direction of Shinjuku; I and a classmate walked in the other direction towards the Wakamatsu-cho stop of the streetcar, as both of us lived close to that stop. It was about two o'clock in the afternoon, too early to go home, and we asked ourselves what to do next, when the classmate suggested our going to the movies. I considered it odd, and I even didn't think such was possible. But I followed him and indeed found a movie theater nearby, which we entered. There were only two or three besides us.

As we waited for the movie to begin, I unexpectedly heard a popular song played over the loudspeaker:

> Walking the street of Yushima,
> Everybody recalls the passionate love between
>     O'Tsuta and Chikara,
> Does the white plum keep it in remembrance?
> Only their shadows remain on the shrine fence.

It was titled "The White Plum of Yushima," and based on a well-known novel of 1907 by Kyoka Izumi. I was somehow familiar with the song, but became anxious, as it did not belong to the type of war songs the government approved. In fact, at that time songs such as the ad-balloon or the Odakyu line had been prohibited under censorship. But of course it was illogical for a labor-mobilization dodger to think that way.

The movie was a slapstick comedy about the war effort of a station master played by a popular comedian. I don't remember anything about the movie other than that, but never forgot the song and my listening to it in such an unexpected place and at an unexpected time, as the melody had a strange resonance with my emotions and my situation from which I could not escape.

The air raid of May 25 destroyed that theater. The sweet and astringent persimmon trees of my childhood were burned, and the Chinese restaurant Ko-Ran was too, so that my wish of eating there was never fulfilled.

During the period between the two raids on April 13 and May 25, my family lived in a Buddhist temple, where our family tombs had been moved from the temple that originally had the cemetery where my ancestors were buried. The head priest, his wife and daughter, and a sexton were living in this temple, which was spacious and had many rooms. The daughter was married and in her late twenties. I think her husband had been drafted, but I did not know where he was.

This married daughter would invite me into her room and would make me listen to her piano playing. The music piece was not a Chopin or a Mozart; it was a tune in a languid mood, perhaps one of those popular in the United States in the 1920s or early 1930s. I think she explained to me what it was, but I don't remember her explanation. In any case it was not of the type that would make the army happy. Indeed, around the end of 1943 the government specified about one thousand music pieces of English and American origin, whose public playing would be prohibited.

It may be meaningless to ask why she was playing that kind of music. Perhaps she was recalling her happier days, as I was doing the same with my childhood. Or she wanted to taste moments of escape from the depressing days with no end in sight.

While the daughter was playing such piano pieces, the father had a guest, an army officer, who wanted to learn something about the Wisdom Sutra, the shortest of all sutras. The young man would come to the temple in the evening and have a session with the priest. I would hear their reading the sutra together, which I recognized, as I knew the words. Of course that didn't help anybody. May I say, they knew it themselves, but it was the time when nothing else would have helped.

In fact, it was in November or December in 1944 that we ordinary citizens lost hope in the war. After that we were dragged by the momentum that had accumulated, pretending that some miraculous event would save us. By the spring of 1945 we had passed the critical point, at least from my personal viewpoint, and we could not but give ourselves to the hands of fate. There was no use in analyzing or grieving at the situation. We just tried to live day by day with resignation. Speaking about myself, what sustained me in this period was my youth and desire for learning. I trusted that my desire would be gratified someday, though I didn't know how and when that day would arrive.

## 7. About Death

The corpses I saw on Ohkubo Dori deeply impressed me with the thought that I might become one of them. Indeed, the probability of my falling into that fate was not small, but I had no fear of death.

My grandfather Kintaro died in 1934, and I remember seeing his face in the coffin, which was the first time I saw a dead person. The teachers in elementary school told us, often emphatically, of the mortality of everybody, but I don't remember in what context they told us so. I was nine years old when the cat we kept died. I was unable to fight back my tears and said, "I'm going to cry," to which one of my sisters said, "Don't be silly."

As a third- or fourth-grader I suffered from tonsilitis and stayed in bed for a week. I didn't expect to die, of course, but while I was watching the fresh leafage of maples out of the window, I imagined, under the influence of high fever, that such beautiful green might be in my eyes when I die. Such a sentimental feeling is common to boys of that age, I think. But the experiences I had at age ten may not be so universal. Anyway here is what happened.

I was alone in the house, and thinking about my mortality, for what reason I don't remember. Suddenly I realized that everything would keep existing after my death. The people would still be walking on the streets under the blue sky as before; only I would not be there. In other words, I knew at that moment that there would be a world in which I did not exist.

This revelation of mine is of course trivial, as I did not exist before I was born. But I thought then that I discovered something important that I didn't know before. I kept thinking for a while about the world in which I did not exist. Strangely that gradually led me into a state of ecstasy. I eventually awoke from that state, and that was that. After a few days, I was alone again, and thinking of the same thing, which produced the same state of ecstasy. I cannot explain what that state was; I only remember that I fell into that state twice. When I tried it for the third time, I had unpleasant feelings, and even had a headache. After that I never did this kind of experiment again, and I never told these experiences to anybody.

I can disregard that ecstasy, but the thought about the existence of the world without myself remained in my mind as a kind of enlightenment. Also, this concerns only me, and so I have no intention of generalization.

Many years later when I was working on a mathematical project, I occasionally had the feeling that I would not like to die at that moment. To explain the meaning of this sentiment,

I must first note that every mathematician makes mistakes, and I am no exception. I would realize that I had been wrong in some of my arguments, but I knew that I could fix them and also that it would take a considerable amount of time. I would have hated to have died at that stage. This is merely a matter of sentiment, as I was not facing death. This is different from, "I cannot die before finishing the work."

I never had any religious faith, though I used to read many religious books. I was merely interested in knowing what the books said; I was never tempted to commit myself to any particular religion.

Some of my classmates died young of tuberculosis. I have been healthy enough to escape that fate, and even lucky enough to survive the air raids, which makes me think that I must live cherishing myself. Or, that I am still living now is enough of a wonderful thing, for which I must be thankful, and I should not ask for more.

## 8. How I Studied

Returning to the last months of the war, I now describe in what kind of intellectual activities I engaged myself. As I already wrote, the student labor mobilization forced us to work at various places from sometime in November, 1944 through August 15, 1945 when the war ended. I did not go to school in that period, but that deprivation gave me a strong desire for learning, much stronger than what I would have had under normal circumstances. I think this applied to many of my classmates too.

We tried to learn as much as possible in our spare time. I read English grammar and physics textbooks but mathematics was the subject of my central interest. Why mathematics? I liked the subject and thought that I would be able to learn many new fascinating things, more so than any other branch of study. I instinctively knew that mathematics was a creative endeavor, not a discipline of expounding established theories, and that suited me. I acquired a good amount of mathematical knowledge since then until the end of high school (in the old system), a period of roughly four and a half years, and it is this knowledge that formed the nucleus of my education as a mathematician.

I had no teacher or mentor; also, it was difficult to find good textbooks. Naturally my methods of learning were inefficient. I was paid as a labor-mobilized student, but there weren't many good books to buy, or rather, practically none. It may be added that our pay was so low that the owner of the factory where we worked made good profits.

At the end of 1944, however, I was able to buy a newly published book on elementary calculus, which I think was intended for engineers. The treatment was deliberately intuitive and not rigorous. For instance, the author determined the power series expansion of a trigonometric function by assuming the function to have an expansion. I think that type of approach is best at the introductory level; I liked it anyway. However, I lent it to a classmate who failed to save it from an air raid. I was angry for his casual attitude towards such a precious object, but had nothing to do but blame myself for trusting that kind of man.

Still I managed to find something useful. One of the calculus books I had in the last months of the war was written in Japanese in the first third, then in English in the remainder. I remember that the switch from Japanese to English was abruptly made in the middle of the proof of a theorem. Even so, the book served the purpose. The task of finding textbooks became easier after entering high school in 1946. Since that time, my mathematical knowledge has been gained from reading books and papers, or from what I did on my own, with few exceptions. In other words, I learned little from lectures. In any case, my life as a student of mathematics in this period, if unorthodox, was ample, far more so than the three years at the University of Tokyo.

The war ended on August 15, 1945. The summer would soon end too. Though the end of the summer in an ordinary year might have suggested a certain poetical sentiment, what we had in the city that year was the end of the summer in the fire-ravaged streets. But we were able to feel differently in the suburbs, where war's scars were less noticeable. I was living in an apartment in the Mitaka district, which had wide fields of ripening corn, pumpkins, and sweet potatoes. Under a gentle breeze that indicated the beginning of autumn, I would walk a path in such a field, and would fully feel that the war had at last ended.

I also have a distinct memory of an incident in the same period which would have been utterly ordinary in any normal time. One Sunday, I was in our apartment and heard the voices of children playing on the street. They were singing a simple song that accompanied their game. The song was familiar to me, but I had not heard it for a long time, and at that moment I was emotionally struck by the sense that the time had finally returned to peaceful old days. I could also hope that things would improve.

This didn't mean that we began living an easier life immediately. That we were able to sleep without fearing air raids was a great change, of course. Two years later, in September 1947, memories of air raids still vivid and finding reasonable housing still difficult, the embankment of the Tone river collapsed at Kurihashi, and a large western section of Tokyo metropolis was flooded. Thus some of those who escaped the fires suffered from a water calamity.

There had always been food shortages during the war, but the matter became worse afterwards. The worst period was from December 1945 until February 1946. We were literally hungry all the time. Even after that, an unpleasant situation, if not as bad as those three months, lingered for almost three years. It was the period of the highest rate of inflation in Japanese history; the government bonds that we were forced to buy during the war came to nothing, and black market merchants became richer and exercised great influence over everything.

I was one of the poorer people, but at least I had the enjoy-ment of learning without being forced to work. In a short period in the fall of 1945 I was living in Tachikawa, an unsophisticated city fifteen miles west of Shinjuku. One day at a second-hand bookstore in the city, I found a copy of *The Sketch Book* by Wash-ington Irving, in a good binding. As the rate of inflation was still modest at that time, the price was reasonable, and I took it to the store-keeper, who said in a surprised tone, "Oh, you've found a wonderful book." It was the time when such books in English had begun to be in demand.

However, to my later regret, I lost the book for reasons un-known to me. I remember reading "Rip Van Winkle," but nothing else. Looking at that work of Irving now, I am puzzled: was I sufficiently proficient in English to be able to read the whole book smoothly? Unlikely. Or it may have been an anthology of short stories by several authors. After that I moved three times before 1957, and each time I disposed of unnecessary books, including some on mathematics, and so it is quite possible that the book got mixed with others; also, I might have given it to somebody. In any case I now think that I should have kept it as a memento of those days which were exciting in a surreal way.

In that period I would have called myself a boy consumed by a strong desire for learning. I thought I would learn more things that I didn't know before, and many interesting things would be awaiting me. I especially felt that way when I enrolled in high school, which was in 1946.

## 9. Evil Ambition and Arrogance

At that time Japanese high school was from the eleventh through thirteenth grades; after that college education was for three years. At present the system is very similar to the American one. Also the Japanese academic year was, and still is, from April to March. Thus my entrance examination of Dai-Ichi Koto Gakko (The First High School) was held sometime in March 1946. In those days there were only thirty-two such schools, public and private, attended by a select few. In addition, some private colleges had their own preparatory schools, whose functions were similar to those of high schools.

Many of the applicants to that school that year were those who had been educated in military schools of various kinds, and the government, concerned about the consequences of admitting many such ex-military students, delayed making decisions regarding admissions. I don't know how they settled the issue, but I was accepted after all.

All students of that school were required to live in dormitories. There were continuing problems with food supply, and naturally we lived on rations. After several weeks of dormitory life, we had a week of holidays for that reason. Before going home, butter released from the U.S. occupation forces, a ten-pound can per two students, was distributed. We somehow managed to cut each can in half, and even managed to consume five pounds of butter, though I don't remember how that was done.

We even got rationed sake, Japanese rice wine. Many, perhaps most, of the students in the whole school were at or above the legal drinking age of twenty. But the school official decided, for expediency, to give each student 180 milliliters of sake, no matter how old he was. I was seventeen or eighteen at that time, and was not particularly thrilled by the availability of sake; I merely said to myself, "Why not?" and went to the dormitory kitchen, bringing my lunch box as a vessel. I received sake in my box, but the situation resulted in the problem of how to consume what I got. Besides, it was awkward to walk back to my room with that box. Without a better idea, I drank that quantity of sake in one gulp on the spot.

Interestingly, it had no effect on me; maybe it was very weak. In those days, there were a few drinkers among my classmates, but I was not among them, nor had I any desire for alcoholic beverages, though I knew that my body was sake-tolerant.

Every Japanese, young or old, drinks a small quantity of *toso*, sake with herbs, on the first three mornings of January, which is an age-old and supposedly salutary ritual. Therefore we, as school children, as well as the school principal, after drinking toso, attended the New Year ceremony I described before, and sang the national anthem under that condition.

Coming to the sober aspect of high school, there was one thing which impressed me on the first day. The head of the school was Teiyu Amano, who became the Education Minister some years later. He talked to us freshmen following customary practice, but I don't remember his talk except for one line, "You can be proud of yourselves and must have self-confidence, as you have been chosen from the very best of your generation." It was true that being a student in that school was highly prestigious, but I was surprised to hear such an explicit incitement without reservation, as I was expecting something like, "Don't be conceited, as there are many who were unlucky but as good as you." So I thought he was really a sensible man. Later I read many articles written about him after his death, none of which contradicted this first impression of mine. He studied German philosophy, but to be an educator was his true vocation.

I think it is important to encourage or inspire young people in a proper way, and he knew it. As to this kind of idea I remember an interesting passage in *Shobo-Genzo-Zuimonki,* which is the analects of Dogen, a Zen monk in the thirteenth century, recorded by one of his disciples. Here are excerpts of the passage in question. (The subject "I" means Dogen himself.)

"After studying at the temple of Eizan, I came to Kenninji [a Buddhist temple at the banks of the Kamo River in Kyoto]. While in Eizan I was unable to find good teachers who could teach me what was right, and so I began to have an evil ambition. My teachers at Eizan encouraged me to become a priest whose name would be known widely to the public, and so I set my aim to be a man as good as those who had won the title of Daishi, Great Teacher, and I strived for that end. But reading two well-known Chinese biographies of famous priests, I found that they differed fundamentally from what my teachers said, and realized that I must change my attitude. Even speaking about reputation, I should not care about the compliments of low-level contemporaries; instead, I should compare myself with those on the highest level in history, and attempt to reach that level. After having this revelation, I began viewing those Daishi as trash, and I then had new thoughts and attitudes completely different from those I had had before."

This is Dogen's recollection of his thoughts spanning several years after the age of thirteen, told to one of his desciples when he was about thirty-six. We can easily recognize in this traces of his thoughts in later years, not necessarily those of his younger days. There were four well-known priests who had won the title of Daishi before Dogen's time. By the time of this recollection, he had formed the idea that the teachings of those Daishi were wrong, and what he said above certainly includes this sentiment, and therefore we cannot accept everything as his thoughts formed in his youth. But we can disregard that aspect, as I am mainly interested in the earlier part of his recollection, which can be taken at face value.

His teachers told him, "Study hard, and try to achieve the status of a famous man who will eventually be called Daishi." Yielding to that incitement, he had an "evil ambition," but he was able to find what was right and headed in the right direction, aiming at a higher or the highest ideal.

Therefore if one rouses somebody, it must be done in the right way. Also, I find "the compliments of low-level contemporaries" to be a pertinent expression applicable even to some present-day situations. Why can't we say Nobel prizes are just that?

Speaking about myself, I never had an evil ambition as Dogen did. I say it with no reservation. I never wished to become a great scholar whose name would be widely known in the world; I never had even one-tenth of that kind of wish. Besides, nobody exhorted me as Dogen's teachers did.

To explain that point in a different way, let me now narrate an incident that happened when I entered the University of Tokyo in 1949. There was a written entrance examination and the names of those accepted were posted on billboards on the campus. On the specified day I went to see the result on the billboard at the Department of Mathematics, where I met one of my classmates from middle and high school, who was accepted into the Department of Agriculture. He said, "Congratulations, but what's the point in entering the math department?" Taken aback, I said nothing. Then he added, "Oh, I now understand, perhaps you intend to become a professor, don't you?" I didn't answer that question either; I just smiled, congratulated him, and talked about other friends of ours who were accepted. But I was thinking, "What? Professor? To hell with it!"

I wanted to satisfy my intellectual curiosity, and I thought I would be able to do something in mathematics, and that was all. Of course I knew that I would have to earn a living, but I thought that would be taken care of without any plan. I became a professor as he predicted, which happened to be so not because of my desire at that time, but because of the development of various events in my life. There were people who helped me in my career. Thanks to them, or because of them, I ended up in my present position.

Let me insert here an incident that happened a few days after the above conversation. One of my high school classmates who was also one of my roommates at the dormitory invited me to spend a few days at his home. It was located in a coastal town in Shizuoka Prefecture, facing the Pacific Ocean, about eighty miles west of Tokyo. He had been accepted at the mathematics department of a state university different from mine. One day we were strolling along the beach, when we met a fisherman who knew him. The man asked us which department we were in. Told that we both were in mathematics, he said, "That's wonderful. You will eventually have jobs in the Finance Ministry, and become high officials of the government." He was not completely off the mark, as we both became government employees as teachers. Besides, one of my Ph.D. students now holds a high position in the U.S. Social Security Administration, and his Japanese counterpart was a math major.

As to Dogen, there is one more story about him which causes me to compare myself with him. After staying in Kenninji for some years, he realized that he would never be able to learn true Buddhism in Japan, and went to China, which was in the Song period. He visited many temples there and met many priests who were supposed to be the most learned and reputed at the time. But he had difficulties in finding a person who was good enough to be his teacher. Two years later, he eventually found one, but before that he formed an arrogant idea that there was nobody better than he in both Japan and China, and he intended to return home. This is recorded in one manuscript version of his old biography, but not in the officially sanctioned version. Therefore it is believable that this passage reflects faithfully what he confessed to his students, and later authors of his school tried to suppress it in order not to make him look like a conceited man.

Speaking again about myself, I never had such an arrogant thought. I remember only one kind of feeling I had around 1963–1964 which is different from, but has a certain similarity to, the arrogance Dogen once had. I came to the States in 1962 for the second time. In the following two years I saw many mathematicians at many different places. They were certainly smart and competent, but did not seem to have great ideas. Indeed, there were several incidents which made me think so. I participated in a conference at Boulder, Colorado in 1963, and also another at Woods Hole, where I almost drowned, in 1964. As I narrate in a later section, I was able to explain many things to the participants of the conferences, which were new to them.

Sometime in 1963 I gave a colloquium talk at the Massachusetts Institute of Technology. John Tate was at Harvard at that time, and I had already met him in Paris. He called me to his office and asked me some questions about certain types of abelian varieties. I answered them and explained my ideas. Later one of his students wrote his Ph. D. thesis on that subject based on my ideas, and published it, acknowledging Tate's help, but not that I helped him. Once in a party in the same period I met Lars Ahlfors, and had a friendly conversation. But at the very beginning he told me that he knew nothing about analytic functions of many variables. Apparently he knew what I was doing, and was trying to avoid an awkward situation, which was all right, but I found it odd.

In any case I thought that at least in my field I would learn nothing from those in the States at that time. I don't think this was my arrogance, as it was really so, and I was merely making my judgment calmly only about the subject on which I was working.

I had no evil ambition, nor was I arrogant, but of course I am not saying that I am better than Dogen. I am simply telling that I was so. As to Dogen, I am truly impressed by his thoughtful words in his analects, and astonished by the fact that he was only in his mid-thirties, at which age I was far less mature.

I would like to insert here something related to another aspect of the passage I quoted from his analects, unrelated to his evil ambition. First I have to mention a well-known joke: Once an Egyptologist deciphered a papyrus document. It said, "Young people these days don't behave." In other words, generation after generation for thousands of years humankind kept saying, "Young people these days … ." But this is not just a fiction. There were many kinds of vicissitudes in religion in the long history of Egypt; for example, polytheism became monotheism at some point, and it changed back into polytheism. Therefore, it is said, the papyrus document recorded the grief of a priest who thought that the religion of his time was corrupt and different from his ideal, and later it was made into a joke. This may very well be true. If so, it may have been, "Old people these days are corrupt."

That is exactly what Dogen said. Although I did not have an evil ambition, in my twenties and thirties I was thinking, "Old mathematicians in Japan these days understand nothing." I would think many of my generation thought so too.

Returning to the point of my entering high school, I now narrate a story which can be taken as an example of "old mathematicians." There were two courses in mathematics in the first year, one of which used to be analytical geometry, mainly concerned with conic sections. Akira Okada was in charge of it, and started the course with an introduction to set theory. This by itself was not unusual, as one of the other teachers was nicknamed *Menge*, German word for set, who used to start calculus courses with the definition of a set, using that German word. As for Okada's course, after a few hours of set theory, he developed an axiomatic theory of two-dimensional Euclidean geometry. At that time Shokichi Iyanaga, a professor at the University of Tokyo, was lecturing on exactly that topic at the Department of Mathematics. Okada had a great respect for Iyanaga, learned the theory from him, and explained it to us.

Probably what Okada did was better than the old-fashioned analytical geometry, but there were three problematic points.

1. The significance of such an axiomatic treatment of Euclidean geometry as a course subject in high school.

2. The significance of such an axiomatic formulation of Euclidean geometry, irrespective of whether it is worthy of teaching.

3. That Okada did not fully understand the theory.

The last point is the simplest. I had that impression, because one of the problems in the written examination at the end of the term was not rigorously formulated. I would call him a man of convictions, who thought he understood something without really understanding it. He was earnest and unique, however.

Points 2 and 3 are mutually related. As every mathematician knows, Hilbert published his *Foundation of Geometry* in 1899, in which he showed that the geometry of three-dimensional Euclidean space can be rigorously developed from a set of clearly stated axioms. It was an epoch-making work, as the old-fashioned Euclidean geometry relied on incomplete and ambiguous axioms. Iyanaga thought this to be extremely important, and even viewed it as a model of true mathematics; in other words, mathematics for Iyanaga was to do something similar to that work of Hilbert.

One can perhaps argue that Hilbert did something which must be done at least once. However, it is also true that Euclidean geometry existed and was well understood without his system of axioms, and his theory added nothing new that we did not already know. He merely showed that what we had known could be developed from the axioms. Going to the extreme, one may say that no mathematician today would be in trouble, even if Hilbert's theory had not been published. Indeed, at present, linear algebra is taught as a standard course subject, but no instructor ever mentions Hilbert's axioms.

However, there are those who think that anything with the name Hilbert attached must be taken very seriously. Iyanaga may be an extreme case, but in any case he had a kind of mathematical faith, and he taught based on that faith. Worse yet, Okada followed him, and forced the wrong idea on us. Speaking about myself, I never had any interest in such an axiomatic approach; the same can be said about my classmates at the University of Tokyo. Iyanaga was certainly among those whom we viewed as "old mathematicians these days," but there were many more.

Though Okada's course did no great harm to us, there is a definite reason for my describing it. Suppose we have a mathematical object A and we easily see that A satisfies a certain set of properties P. It often happens that it is natural to conjecture that a mathematical object with properties P must be A, but it is difficult to prove it. Then many mathematicians consider the

solution of that problem a great achievement. I said "many," but
"almost all" is often more appropriate. I think this type of thought
is fundamentally wrong.

There are many cases. If A is a collection of many objects with
the said properties, the situation is different. If A is absolutely a
single object, then the solution adds no new knowledge, as in the
case of Hilbert's foundation of geometry. The matter would be
different if there were some interesting applications, but in most
cases we find no applications. In any case, it is absolutely wrong
to think "the more difficult, the better" in mathematics. If one
proves a theorem, its importance depends solely on the theorem,
not on how difficult the proof is. But strangely, there are many,
incredibly many, mathematicians who say, "That theorem is not
of much value, because its proof is not difficult."

Returning to Hilbert, we know of course that he did many great
things other than the foundation of geometry. It must be pointed
out, however, that he was not familiar with all branches of mathematics. For example, his understanding of the theory of quadratic
forms was superficial. He did not appreciate properly, or rather,
never tried to understand, the important work of Minkowski, two
years his junior, on this topic. In addition, his taste in geometry,
generally speaking, was bad, and he posed minor problems whose
solutions would produce no major development.

I already mentioned non-Euclidean geometry in connection
with spherical trigonometry. Hilbert was of course aware of low-
dimensional cases of non-Euclidean geometry, but had no interest
in the higher-dimensional case. The theory of symmetric spaces,
which can be viewed as a natural extension of non-Euclidean geometry, was developed by Elie Cartan in the 1920s, and one might
say, it would be too harsh to criticize Hilbert for his lack of in-
sight in that direction. But the Erlangen program of Klein in 1872
already included special cases of symmetric spaces, a fact Cartan acknowledged. Therefore Cartan could have included Hilbert
among "old mathematicians these days" in that sense, though it

must be noted that Hilbert was not yet forty when he posed his *Problems*.

I later tell one more story about the futility of the axiomatic approach. Changing the subject for the moment, let me now narrate my learning foreign languages in high school. I studied English and German in high school. My teachers in German were Hidehiro Hikami and Michio Takeyama, both well-known scholars. The latter especially was known as the author of the novel *Harp of Burma* (*Biruma no Tategoto*). I read *Der grüne Heinrich* by Keller with him. Incidentally he graduated from the same middle and high schools as I did. In 1990 I underwent an operation at a hospital in Princeton, and my anesthetist told me that he had read the novel, and was very impressed to hear that the author had been my German teacher in high school.

As to Hikami's class, I remember the following. The roll call was standard, and answering for another was common. At that time he would teach two consecutive forty-five minute classes, with a fifteen minute recess in between. One day I answered for a classmate at the beginning, but he attended the second class. Then Hikami called his name to read the textbook, expecting that I would be the person, but to his surprise, another person stood up. The teacher, much confused, said to me, "Aren't you Mr. So and So?" to which I said, "No," and the person himself said, "It's me," and that was that. I never knew whether he was able to solve the mystery; maybe he thought we were playing a game on him for fun.

After learning German for a month or two, I was able to read German books on mathematics. I think I first tested my knowledge of German with Bieberbach's book on differential equations with no intention of reading the whole book; I don't remember why I picked it. After that, while in high school, I read several German books including the first volume of *Moderne Algebra* by van der Waerden. As for French, I read a small book on French grammar, and later went to a school specializing in French, but quit after a short period, when I became capable of passably reading French books on mathematics. I realized that my French must be improved when I planned to go to France, but I will come to that later.

There was one more foreign language I learned: Russian. But it was around 1955, when I was teaching at the University of Tokyo. One of the Japanese translators of Hogben's book, Saburo Yamazaki, was a senior member of the same department, and very good at Russian. He said he would teach Russian to us younger people, saying that it was easy to learn Russian for someone who already knew German. Indeed, he did teach us using lunch time while we were eating. Each of us got a Russian textbook on linear algebra, and tried to read it from the beginning. He would read the text, and let us imitate him; he would then explain grammar. I acquired a grammar book too. He was certainly a good teacher, and after several sessions, I was able to read Russian papers in mathematics with the help of a dictionary.

My knowledge of Russian was put to use after three years in an unexpected fashion, when I was a member of the Institute for Advanced Study. One day André Weil found a Russian paper in the latest issue of some Russian mathematical journal, which he considered interesting and worthy of circulating among the members, almost all of whom were unable to read Russian. Strangely he came to me and said, "Can you make an English translation?" to which I answered, "I'll try." I believe he inherited Russian blood from his ancestors, and was actually fluent in Russian. But it was typical of him to make somebody else to do a job which he could do himself. Anyway I made an English translation, and copies were handed out to the interested members. Even today I am puzzled: Why did he choose me for that task?

Some thirty years later, I advised a senior at Princeton, who emigrated from Russia. I chose a German paper for the topic of his senior thesis, but he was unable to read German. So I told him that it would be advantageous for him to learn German at that stage, as there were many important German papers, and also that he should be able to learn the language quickly, as it had similarities with Russian. He was reluctant at first, but eventually he worked as I hoped, with a happy ending.

Returning to my high school years, the students were eager to read some standard literature, which was supposed to be beneficial to character building: novels, dramas, essays, philosophy or whatever they thought intellectually stimulating. Jiro Abe, Eijiro Kawai, and Hyakuzo Kurata were popular names; also, Hermann Hesse, Jean Paul Sartre, and Alain were often mentioned. The classmate of mine who was raised in the States was passionately reading *Jean-Christophe* by Romain Rolland. Strangely, history books were absent from their list. In any case, I was attracted to no such writings. I didn't think much of such cultural training by reading. Also, to confess, many mathematicians, including myself, are secretly contemptuous of most philosophical writings by saying, "They are just rhetoric; no theorem is proved."

Many years later I read Kawai's diary, and found him to be an interesting man. He was a type whose actions almost never followed his words. His diary described his daily life filled with earthly and ignoble desires. But there was something commendable about his political position, and I thought that the showdown between him and the hypocritical Marxist economists who exercised great influence in post-war Japan would have been very interesting, but regrettably he had died during the war.

It is difficult to tell other people what one is reading, as it always causes misunderstanding. Shigeru Yoshida, one of the best-known prime ministers of Japan after the war, was once asked what he was doing. It was several years after his retirement. He said, "I am reading *Torimono-cho*," to which the Japanese public reacted with disapproval and disappointment, as they wished him to be more sophisticated and in good taste. *Torimono-cho* may be translated as a detective's casebook, but the setting is in the Edo period, and apparently it was viewed as too lowbrow. But actually, some of the stories are quite good and not vulgar. If he had said, "I am rereading *the Decline and Fall of the Roman Empire*," that would have been pretentious. I think what he said really made sense.

As for myself, I mention here one book I read in my childhood, not in high school. There was a colloquial translation of *Ochikubo Monogatari* (*The Tale of Lady Ochikubo*), a novel written in the tenth century about forty years earlier than *The Tale of Genji*. The book, published in 1912, was intended for both adults and children, and had several full-page illustrations in color. Our family had a copy, which I read as a child. It became my favorite, but was destroyed by the air raids. Later, I read the novel repeatedly in its original version. The stories are narrated in a vivid realistic style, rarely seen in Japanese literature of that period. I think this is one of the very best classical Japanese novels. Besides, such novels in the modern style appeared in Europe only several hundred years later. But clearly it was not the type of literary work discussed among high school students.

## 10. Three Years at the University of Tokyo

My wish that I would be able to learn plenty of good mathematics at the university was soon betrayed by reality. For one thing, while in high school, I had acquired a decent amount of mathematical knowledge, and there was not much new in what was being taught in the first year at the university. But more importantly, the professors and associate professors at that time did not give much serious thought to the question of what should be taught. It was the same story as what I experienced in the first year in middle school. They were simply repeating the old stuff, which should have been replaced by better material. Not that fashionable or ultra-advanced theories must be taught. One has to think carefully even at elementary levels. I already mentioned the stupidity of teaching the crane-tortoise problem, but the same type of stupidity existed at the university level. My teacher in watercolor who spoke futurism, was far more inspiring than those university professors.

I will now describe what Kenkichi Iwasawa taught, though it was one of the better courses. He was well known for the theory named after him. Four years earlier in 1945, Artin and Whaples had jointly published a paper, and Iwasawa's main objective was to explain the contents of that paper. He began with an introduction to valuation theory, and then turned to the paper; the whole term was expended for that purpose. The part concerning valuation theory roughly corresponded to the portion of his book, in Japanese, on algebraic functions of one variable, which he was then preparing. The lectures were clear, and not perfunctory, but there was much to be desired.

He had no intention of explaining the basic ideas to the beginners. There were no examples: just definitions, theorems, and proofs. In other words, he was lecturing for himself, not for the students.

Putting that point aside, the larger problem was that the original paper of Artin and Whaples was insignificant. It was similar to Hilbert's foundation of geometry. Without going into details, suffice it to say that if we start with a set of axioms, then we can show that the objects satisfying the axioms are exactly what we already know, and nothing else.

Thirty-two at that time, Iwasawa was searching for new ideas, perhaps through the process of trial and error. His mathematical knowledge was not very broad, but at least broader than all of his colleagues, as he was familiar with Lie groups in addition to algebraic number theory. His book on algebraic functions just mentioned was a good introductory book at that time. However, he had not yet found any idea that could really be called his own, and he did so more than five years later.

Actually, there was an excellent topic he could have chosen instead of that paper by Artin and Whaples. He presented a short paper at the International Congress of Mathematicians in 1950 held at Chicago. He showed in it that the analytic properties of the Hecke $L$-functions can be derived by using adeles and ideles, as it is ordinarily done today. Therefore after valuation theory, he could have explained his ideas on this subject. Of course it would have been too much to develop the theory rigorously, but I think it was quite feasible for him to explain the basic ideas, which would have made an extremely interesting and inspiring course. But alas, he was more interested in learning the contents of someone else's paper, than in guiding students in the right direction.

The insignificance of the Artin–Whaples paper became self-evident no more than a few years later, possibly even at the time of its publication. Nobody in my generation paid any serious attention to that work. Artin visited Japan in 1955, but we younger people were not interested in what he said, as we knew that he was a has-been and had no interesting new ideas; he was being surrounded only by "old people those days in Japan."

All other courses were just traditional; no professors knew the necessity of teaching new material, and so I did not take their courses seriously. I only tried to get course credits sufficient for graduation. Let me describe how the courses on complex analysis and real analysis went. The former, though competently taught, was not particularly interesting. The professor would call on a student to solve each one of the problems posed a week earlier. One day, somehow many of my classmates were in the classroom ten minutes earlier than the specified time, and so we decided to write down all the solutions on the blackboard, without waiting for him. When that was finished, he arrived, and was naturally surprised, but he simply checked and accepted our solutions. We never repeated it, but possibly that gave him the impression that our class was quite receptive, and he taught us more topics than usual. Indeed, he ended his course with an introduction to value distribution theory, and one of the problems in the final examination was on that topic.

As for real analysis, an associate professor was developing a very general measure theory, and even published a thick book on that topic around the same time. His course was a part of that book. He would boast in class, "You may not be able to realize it, but in fact, after twenty years, many mathematicians in the world will begin studying my theory." However, his final examination concerned only the theory of Lebesgue integration on the real line. He said, "You only need to study the material in my book;" he had just published a book on that subject, different from the thick one. So I bought a copy, priced 250 yen, as I had to familiarize myself with his peculiar terminology.

The exam consisted of four problems. As to one of them he said, "This problem is too difficult for you; I will give you a hint." Every student who finished the problems was supposed to bring his solutions to his office. I did so, but I did not use his hint, which displeased him. He tried hard to find mistakes, but to his disappointment, he found none, and said, "I'm not really convinced, but today I tolerate you." After this I immediately sold the book to a second-hand book dealer, regaining 130 yen, as my copy looked brand-new. Some years later I learned that this professor gave a scroll to one of his students as a certificate of his sanction. That much was not objectionable, but I also heard an unprintable tale about him.

Another associate professor was giving a course on differential geometry. He was reading a second-rate textbook written by an obscure European mathematician, and his lectures were his translation of the book into Japanese, and nothing else. However, he gave a course credit to any student who requested it, with a single grade "good." At that time he was contributing many short essays to newspapers and popular magazines, and so I think in the 1950s he was the best-known mathematician in Japan.

These tell roughly how I was doing with the courses, and otherwise I was reading books on my own; for instance, I read Chevalley's *Theory of Lie Groups*, which was extremely beneficial, and

which I think is far more significant than any book by Hermann
Weyl, at least to me. In addition to reading books, I was trying
various small things on my own. For example, I was able to find
new proofs for the existence theorem for the solutions of a system
of partial differential equations employed in Chevalley's book. I
describe here two more topics which concern infinite sequences of
integers, and which, I hope, the reader will find interesting.

The first topic deals with the powers of a fixed positive integer,
say 3. Arrange the powers as follows:

$$3^1 = 3,$$
$$3^2 = 9,$$
$$3^3 = 27,$$
$$3^4 = 81,$$
$$3^5 = 243,$$
$$3^6 = 729,$$
$$3^7 = 2187,$$
$$3^8 = 6561,$$
$$\cdots = \cdots$$

Here the right-hand sides are arranged so that the first digits form
the first column. Any fixed column gives an infinite sequence of
digits (including 0). Take the second column, for example. Then
we obtain a sequence

(1)        7, 1, 4, 2, 1, 5, 9, 9, 7, 3, 5, 7, 4, 3, ...

Pick a digit $k$, which can be 0. Now the question is: *What is
the probability of $k$ appearing in this sequence?* More precisely,
denote by $f(m)$ the number of times $k$ appears in the first $m$
integers of this sequence. If $k = 5$, for example, then $f(1) =
f(2) = \cdots = f(5) = 0$, $f(6) = \cdots = f(10) = 1$, $f(11) = 2$, and so
on. Then we ask: *Does $\lim_{m \to \infty} f(m)/m$ exist? If so, what is the
limit?* We can indeed determine the limit explicitly. Though the
problem is not difficult, I may be excused to say, "I will give you
a hint," imitating that associate professor. This is a special case

of *uniform distribution.* I am certain that the reader who tries to solve this problem will have a very enjoyable time.

The second topic concerns a polynomial $F(x)$ with integer coefficients. Take

$$F(x) = x^3 + x^2 - 2x - 1,$$

for example. For an integer $n$, we consider the decomposition of $F(n)$ into the product of prime numbers. We can allow $n$ to be negative, but let us assume $n$ to be positive here. Thus

$$F(1) = -1, \ F(2) = 7, \ F(3) = 29, \ F(4) = 71, \ F(5) = 139,$$
$$F(6) = 239, \ F(7) = 13 \cdot 29, \ F(8) = 13 \cdot 43, \ F(9) = 7 \cdot 113, \ \ldots$$

The prime numbers appearing as factors of $F(n)$ form a sequence

(2)               $7, \ 13, \ 29, \ 43, \ 71, \ 113, \ 139, \ 239, \ \ldots$

Now the question is: *What are these prime numbers?* In fact, we can prove that every such prime number $p$, excluding 7, has the property that $p + 1$ or $p - 1$ is divisible by 7. Conversely, every such prime number appears as a factor of $F(n)$ for some positive integer $n$.

While learning class field theory on my own, I realized that the main theorem in easier cases can be formulated in terms of prime factors of $F(n)$ as above, and at that moment I was very happy. The polynomial $F$ cannot be taken arbitrarily. Actually the equation $F(x) = 0$ has $2\cos(2\pi/7)$ as a root, and that fact is essential. If $F(x) = x^2 - a$ with an integer $a$, the problem can be solved by the quadratic reciprocity law. In fact, my later work on the so-called complex multiplication is closely connected with this question of finding $F$ for which the sequence corresponding to (2) can be determined. I should also note that these two problems on infinite sequences of integers are anything but mathematical puzzles; they are serious mathematics. Incidentally, I was never interested in artificial puzzles.

In this way I amused myself, and my time was not entirely wasted. But overall, the three years of college were not a pleasant period. Fortunately, I did have a wonderful day during this period unrelated to mathematics, to which I turn in later sections. However, I note here one important aspect of the political situation around the same time.

The Korean war started in June 1950. In August the North Korean army overwhelmed the United Nations forces, which were squeezed into a small area surrounding Pusan, and the latter's total defeat looked imminent. One day in that period, I ran into a high school classmate, who was also a university student, at a suburban railway station. I said to him, "It looks as if the whole peninsula will soon be in the hands of the communists," to which he replied, "Oh yes, the victory will be gained soon; it is only a matter of a few days patience." To my astonishment, he had a great joy in what I took to be a grievous situation. I thought, "No use in talking to such a guy," and changed the subject of our conversation.

It may sound strange to the reader nowadays, but actually that type of opinion was common to university students, and even prevalent among the so-called progressive intellectuals. Many students were communists or sympathizers, and they had great influence on other students. One of them, a year senior to me in high school and university, would say, "We will have a revolution within five years." I never took such an idea seriously, and I even sensed a selfish and dishonest motive in what they were doing. Therefore I always kept a certain distance from them, but what I heard at the railway station was a shock to me, as I thought the man was moderate.

But many of those much younger than my generation held the faith that communist countries, North Korea, for example, were utopias even around 1975. One such man once said to me, "You are an awfully nice person, and would be nicer had you not been that reactionary." When I reminded him of that comment many years later, he was quite embarrassed and apologized to me. He is now the president of a well-known private university. Also, not a few intellectuals believed that it was the South, not the North, who started or provoked the war. One of them, a certain Odagiri, wrote in an article published in a magazine in 1985 that "I had believed that the South was responsible, but I now realize that it was not so," which of course caused contemptuous comments. It is unbelievable that the man maintained such a faulty idea for thirty-five years, but he was honest enough to admit his mistake. I think there were more of those who maintained the wrong idea and never expressed any remorse for it.

In the 1950s through 1970s many Japanese intellectuals had faith in communism. There were also those who held the opinion that Japan must take the middle ground without becoming closer to the United States or the Soviet Union, which I found patently hypocritical. That was their way of expressing anti-Americanism, and they knew that anti-Americanism appealed to the public, university students in particular, more than anti-Communism. In other words, they were afraid of being viewed as right-wing. I believe such people, the so-called progressive intellectuals, formed the majority of political scientists, economists, and commentators on political affairs, and even some editors of a major newspaper at that time. It is quite possible that they thought that was a clever way of surviving in the world of academia or journalism.

Michio Takeyama, the author of *Harp of Burma,* was completely different. He wrote many articles in which he exposed the evil practices and deceitful behavior of communist countries. Also, he rightfully criticized the International Military Tribunal of the Far East as totally unfair and unjust. He wrote what he thought ought to be written, without fearing how he would be viewed. It is now clear that he was right, and I truly regret that few people nowadays remember how courageous he was and even fewer discuss his achievements.

I should also mention that such progressive intellectuals were happy to overlook many wrongs the Soviet Union did to other countries, especially to Japan. Immediately after the war the

Soviet army detained many Japanese soldiers, brought them to Siberia, and forced them to work for a long period. This mobilization was planned carefully in advance in order to obtain enough labor forces to develop the lands. The soldiers were forced to engage in hard labor, in severe cold, and with meagerly rationed food. I am hoping that the compatriots of my generation feel the same way as I do, and also that the younger generation will recognize anew the significance of this historical event.

In this connection I must note a bizarre incident concerning a course I took in my last year at the university with that boastful associate professor. To get a credit, we had to write something about the topic he gave us, which I did. At that time there were two students who had entered the school a few years earlier than I, but who had not graduated, whom I will call K and G here. One day, K came to me and said, "Both G and I intend to graduate this year. I can take care of myself, but G cannot write the necessary term papers on time. Could you produce one for him for that associate professor's course? G would really appreciate it." Since I had already written one for myself, I did not like the idea much, but I drew up a paper, and handed it to K, believing that G would use it.

Then, on the day of graduation I found the name of K on the list, but not that of G, which naturally puzzled me. But in those days I was so softhearted that I could not guess the reason for it, which must have been obvious to most people. I did find out of course, at least a few months later when strangely K tried to avoid me. I should add that both K and G were among those who had predicted the revolution within five years; I also believe that K once belonged to the Communist Party of Japan. His act was not as hideous as what some of his fellow communists had done, but it did reinforce my conviction that they had no decent moral principles.

Apart from politics, a common form of recreation among university students in those days was mountain-climbing, an activity which may also be so at present. However, the number of climbers

was perhaps smaller than now. In the summer of 1950, two classmates and I ascended a few peaks at least 8600 feet above sea
level, in Nagano Prefecture. In 1951, I went up alone to a peak
in Yamanashi Prefecture of about the same height. I first got off
the train at a railway station, crossed a pass, 4800 feet above sea
level, put up at an inn for the night, and the next day reached the
peak. Then I descended a steep slope, and finally around 8 o'clock
in the evening arrived at a hotel, which was situated at the end of
a ravine called Shosen-kyo, a popular tourist destination. On the
last day I became a sightseer enjoying the splendid scenery of the
rapids. My trip took three full days. I met only two climbers at
the top of the mountain. Otherwise I was all alone on the second
day until I reached the hotel. I saw many tourists on the last day.

This can be counted among my reckless acts, as it was clearly
dangerous to attempt to ascend that type of mountain alone.
Though it was an interesting experience, I resolved to be more
careful after that. One certainly needs to take good care of one-
self, particularly if one has high aspirations.

I had many interesting experiences during and after the war,
which I would never have had in ordinary times. I already mentioned a few of them in earlier sections. Let me add one more here.
I think it was four or five years after the end of the war. One day
my brother asked me to make an electric record player. I don't
remember where he got that idea. Anyway, I got hold of an issue
of a popular science magazine, which explained how to construct
such a player, went to some shops in the Kanda district in the
center of Tokyo, and bought a turn table and a pick-up. Following
the instructions on wiring, I soldered a copper wire to the top of
a vacuum valve in a radio we had at home. Namely I used the
speaker of the radio for the player. I put the turntable on the top
of a wooden box, which was formerly a case for confectioneries.

Finally we had a homemade gramophone. Our next problem
was records. At that time there were many shops in Kanda selling
second-hand records. In fact there were many record companies

in the early 1930s which were selling the likes of the ad-balloon song or sets of Western music, as I noted before, and the quantity of what they produced was enormous. Some of them were of course destroyed by air raids, but many remained intact. My brother made an interesting choice, which was the narration by Hisamoto Shimazu, who was a professor at Tokyo Imperial University. The front side was the passage of *Suma* from *The Tale of Genji,* and the reverse side the passage of *Ohara Goko* from *The Tale of Heike,* each story closing with a Waka poem. Our player worked perfectly. The reading was very cadenced, and we listened to it many times, and so we ended up memorizing all the words as well as the professor's intonation. Incidentally, those were the only passages of the two tales I could recite.

It must be noted, however, that we were not the only people who played the same record over and over. At that time the only broadcasting system was state-owned, and played Western music from the records they had, but their repertoire was very limited. For example, they had nothing by Dvorak other than the *New World* symphony. The only opera number by Puccini was *One fine day.* They tirelessly played Kreisler's encore pieces. I would think some of the music schools and individuals had more, but as for the national broadcasting station, probably they had no funds to acquire new records, even five years after the war.

Apart from the availability of funds, there was another problem, that is, the narrow-minded thinking of the Japanese intellectuals concerning music, art, and literature, who had a propensity to exaggerate the importance of the few they liked, downgrading everything else. I am speaking about 1950s and 60s. For example, Georges Rouault was in fashion, and was talked about more than any other painter. It was so even in 1967, as a twenty-five volume Japanese encyclopedia published that year will attest. His entry includes photos of his works on two full pages, one in color, the other in black and white. Actually he was praised as a unique painter in the 1950s in France; indeed, he was accorded a State funeral in 1958, though that kind of honor is much influenced by other factors such as how old he was when he died. But I found the Japanese enthusiasm for him at that time misguided and unbalanced.

As for literature, Camus, Gide, and Sartre were favorite authors in Japan in the 1950s. As I describe in later sections, I would go to Paris in November 1957, and for that reason I was learning French at l'Institut Franco-Japonais in Tokyo, in 1956 and 1957. Though I read some French literature on my own, for the most part I read the works assigned by the teachers there. For example, I read *L'Étranger* by Camus and *Le Silence de la Mer* by Vercors. One day we had dictation without notice and without explanation as to the source. Later I discovered that it was the first paragraph of *Le Chien jaune* by Georges Simenon. Perhaps the teacher, a Catholic priest, absent a good idea about what to do that day, arbitrarily picked a book he owned, and read its first paragraph to us. He was not a bad teacher, but lacked the character by which he could win our esteem.

The novel *Les Célibataires* by Montherlant was the text of a course taught by a female teacher. An early paragraph described the main character of the novel, a bachelor, who lost many buttons, and fastened his jacket with many clips or safety pins, so that he looked like a man wearing a suit of armor. I was assigned the task of reading it and translating it into Japanese, which I did. After that the teacher asked me, "Could you explain why he was like that?" I found the question silly, and said nothing. Then she said, "Your reading and translation are fine, but that's nothing if you don't understand such a point," and added her explanation. I was forced to enjoy her course in such a queer way.

After some time many of my classmates realized that she was incompetent. But there were excellent teachers; Tadashi Kobayashi, who was an associate professor at the University of Tokyo at that time, was one.

One day in that period, a professor of mathematics, who was proud of his proficiency in French, asked me what kind of French literature I was reading, which happened to be short stories by Anatole France. Hearing my answer, he said, "Anatole France may be all right, but you must read Gide." This may sound normal, but in fact there are two problems with such a statement. First of all, Gide was "the author" of the time. It is similar to "Cézanne may be all right, but you must look at Rouault." There is another point that those of the professor's generation had a tendency of making pompous statements with no content, for the purpose of showing their authority.

At that time I was already teaching at the University of Tokyo, and so I learned a great deal from my experiences as a student in class, and also from such encounters. In particular I learned what I should do and what I should not. There were those who were reluctant to accept the right answer, and were always more interested in nitpicking, like the man who said, "Today I tolerate you." I told myself never to be condescending to somebody in such a way.

There was a maxim by Vauvenargues: "Being unable to have enthusiasm for anything is a sign of mediocrity." I don't know whether this is an accurate quotation. Possibly, he said something more concrete. It is said that he was the type of person who would rather be a great thief like Maurice Leblanc's Arsène Lupin than a good person whose existence nobody notices. Be that as it may, I somehow transformed the maxim into "Your existence doesn't count unless you are able to praise something daringly." Indeed, I have known too many people who always speak in a lukewarm way.

School children boarding a train at Shinjuku Station on their way to a summer school program in Nagano Prefecture, July 1936. For more details, see page 32

The photo of Tomoo Tobari (right) and the author (left). Paris,
March, 1958

The author holding his daughter, Tomoko Shimura.    Mitaka, Tokyo, April 1960

# CHAPTER III

## AS A MATHEMATICIAN

## 11. My Embarkation as a Mathematician

After graduating from the University of Tokyo in 1952, I obtained a position as an assistant at the College of General Education of the University of Tokyo. This was a tenured position in the sense that I was a government employee who would never be fired, and more significantly, would receive an extremely meager salary. All members of the department above the age of twenty-six at that time were content to be teachers of calculus, and I had no intention of becoming a member of that group. In the period 1952–1953, I began to find vaguely in which way I should proceed. I started a definite mathematical project in the early months of 1953, and completed it around October that year, which I can call the starting point of my career as a mathematician. Before narrating this project, I first mention two incidents.

In that year Claude Chevalley visited Japan, and gave a series of lectures at the University of Tokyo on his theory of algebraic groups. He was developing the theory on his own, which mainly concerned linear algebraic groups. His lectures were well prepared, and contained new formulations which were useful for researchers in that field, but as a whole they lacked freshness, at least to me. I was looking for something really new, which would let me see a new horizon.

He proved in his lectures a basic theorem on the function field of a linear algebraic group, which he later published in a

G. Shimura, *The Map of My Life,* doi: 10.1007/978-0-387-79715-1_3,
© Springer Science+Business Media, LLC 2008

paper (Journ. Math. Soc. Japan 6, 1954). He employed a lemma (Lemma 2 of the paper) concerning subfields of a purely transcendental extension of a field, and the lemma by itself was clear-cut and interesting.

I soon found a simpler proof of the lemma, which I think was some time in the spring of 1953; in fact I had proved it in an easier case, which I explain below. My proof was communicated to Chevalley, who substituted it for his original proof, and wrote that "The following proof of this lemma has been communicated to me by Mr. Shimura." His paper attracted the attention of many researchers, and probably this was the first time my name was known to the mathematical public, at least to that extent.

Somewhat earlier, at the end of March of the same year, Yasuo Akizuki, a professor at Kyoto University, organized a small conference at his school, which could be called a get together of younger mathematicians in Tokyo and Kyoto working on algebraic geometry and number theory. I was not among the planned speakers, but at the end Akizuki, who thought there was enough time, said to me, "Why don't you present something?" I was not prepared, but gave an impromptu talk about the proof of an easier case of the lemma mentioned above. Actually I had proved another simple fact, which is well-known nowadays but was new at that time, and Akizuki had known my proof. It is probable that he had that in mind, when he urged me to talk.

Here are the names of those present at the meeting: in Kyoto, Yoshikazu Nakai, Shigeo Nakano, Junichi Igusa, and Hideyuki Matsumura, and possibly Mieo Nishi; from Tokyo, Ichiro Satake, Michio Kuga, and I. Tsuneo Tomagawa was at Tokyo and Masayoshi Nagata at Nagoya, but they were not there. Nakai, at the age of thirty-three was the oldest; everybody else was no older than thirty.

I remember another incident in the same year, perhaps a few months later. It concerns a paper published in the same issue in which Chevalley's paper appeared. The author was two years senior to me at the university, and I was asked to referee the submitted paper. I found it strange, as I was a novice who graduated from the university only in the previous year. In those days there were fewer mathematicians and fewer papers than now, and I guess the editors had difficulty in finding a suitable person familiar with the subject of the paper, and somehow chose me. In any case, I read the paper, and soon discovered a mistake. It was minor, but clearly the paper could not be published without revision. I made a short note that described the mistake, and brought it to the editorial office. The matter was settled in a practical way without any bureaucratic delay, and I even got a letter of thanks from the author.

I was being viewed as a researcher in such small ways, but there was one thing which played a far more definite role in my career. It was the completion of the project I mentioned at the beginning of this chapter. I do not explain it here; I did not gain a new perspective from the work. In order to gain a perspective, I needed to do various things, and one of the absolute necessities was a theory called "Reduction modulo $p$," which was what I had done. I knew the direction in which I was going, and even had a plan, if vague.

I finished the final version of the manuscript some time in December, 1953. Following Iyanaga's advice, I sent it to André Weil, who was in Chicago. His letter, dated December 23, 1953, reads as follows:

Dear Mr. Shimura,

I am very much interested by your manuscript. This is a very important step forward, and will certainly play a basic role in future applications of algebraic geometry to number-theory, which is a subject in which I am deeply interested, as you know. From Deuring's paper on complex multiplication (Math. Ann. 124 (1952), p. 393), you may see how even a few rather elementary results in the reduction-theory (those of Deuring himself, Math. Zeit. 47 (1942), p. 643) are useful in this subject. I have no doubt that your work can be used to simplify and improve greatly the present theory of complex multiplication. It also seems that it supplies just what is needed for new developments in Hecke's theory of modular functions and its

extension to modular functions of several variables; this is a very promising field on which Prof. Eichler is working at present. . . .
. . .

This was the very beginning of my long relationship with him. As to more about the consequences of this letter exchange, the reader is referred to my article "André Weil as I Knew Him" reprinted as Section A5 of this volume. I merely add here that the last comment referred to Eichler's well-known work of 1954, and I eventually proceeded in that direction myself.

I was certainly happy to receive such a letter, but was not dancing for joy, as other young men in my position might have been. I have a hot-headed aspect in my temper, though I show it rarely; also I am easily moved to tears to an embarrassing extent. However, I am incapable of surrendering myself to exultation. That is my nature.

It is told that a Japanese philosopher was able to read through *The Critique of Pure Reason* by Kant, when his mother prepared especially for him cooked rice with red beans, which Japanese eat normally on festival days. I heard similar stories about Japanese mathematicians, but nothing of that sort happened in my life. I never told my family members about my mathematical work. I usually don't tell other people what kind of work I am doing. I know what I am doing, and that is enough. Many years later, my mother was telling her acquaintances about my leaving for France like this: "My son just told me, 'I will leave tomorrow'," which is of course untrue, but contains a grain of truth in the sense that I belong to the type who does not talk about himself.

When I received Weil's letter, I think I showed it naturally to Iyanaga, but otherwise I did not tell anybody else about it, simply because I was a person like that. However, that does not mean that I had no emotional reaction.

At that time I used to walk the western suburbs of Tokyo, which was so even in my days as a university student. Normally my walk was more than an hour, and had no particular destination. There were carved stone slabs dedicated to Seimen Kongo, a

Buddhist demon-god, erected on the roadside in the Edo period. I once made a map that indicated the locations and ages of such slabs. In the spring, I would visit the Tama Cemetery, which had avenues lined with gorgeous cherry trees, and I would bathe in a blizzard of falling cherry blossoms.

Returning to my reaction to Weil's letter, one fine day in January the following year, I took such a walk. Light snow had fallen the previous day. There is a certain warm feeling on such a day after snow in Tokyo. Walking on a country road partly covered by thawing snow under bright sunlight, I had a gentle feeling of satisfaction about my work, and was in a placid state of happiness.

## 12.  As a Teacher

I became a lecturer in 1954 and was promoted to associate professor in 1957, teaching linear algebra and calculus; I once taught an advanced course at the department of mathematics. Every teacher has his or her favorite stories of difficulties in dealing with certain types of students. Here are some such stories of mine, which will endorse my philosophy that there are always those who refuse to be taught. One of the simplest examples concerns a problem:

*Take a coordinate system on a plane, and determine the equation for the straight line passing through the points whose coordinates are* $(3, 5)$ *and* $(3, 8)$.

The answer is $x = 3$. But if one thinks that the equation must always be of the form $y = ax + b$, then one has to divide something by zero, and fails. In fact, I discussed a problem of the same nature in the three-dimensional case when I was teaching, but here I consider the two-dimensional case for simplicity. I would tell students that they should not apply a formula to a problem indiscriminately, and explain this problem. I would even tell them that this would be included in the final exam. But I would later find many answers such as: "I have to divide something by zero, which is impossible, and therefore such a straight line does not exist." Then in the next year, I would repeat everything, and tell them not to make such a silly argument. But then again that type of answer would appear. Clearly they had been taught in middle and high schools that mathematics was merely the art of applying formulas to given problems. I could not change it.

As I narrate later, I moved to Osaka University. There was a student at a state university somewhere, who wanted to change from that school to Osaka University. The official rule said that if there was capacity for admitting more, then the applicant could be admitted provided the person passed an ad hoc examination with a good grade. I was in charge of it and composed a set of four problems one of which was the above question on a straight line. Since he was supposed to have learned the easy part of abstract algebra, I included the following:

*State the definition of a subgroup of a group, and prove that the intersection of two subgroups is also a subgroup.*

This is something one should be able to answer immediately without thinking. I showed my problems to other professors, who felt that the problems were too easy, but I assured them that my choices would be all right. It turned out that the student failed on both problems. The unanimous opinion of the professors was that he could not be admitted, and he was told so. But he left with a parting remark, "I was unfairly entrapped." He could have learned an important lesson, but he was the type of student who would never learn a lesson.

I also recall my conversation with a young Japanese mathematician in the late 1960s or early 1970s. He was a new member of the Institute for Advanced Study. At a party he said to me in a happy tone,

"I was really impressed by the vast number of games sold in this country."

There were no computer games then; what he said referred to the various kinds of board games, jigsaw puzzles, and the like he found at a drugstore in Princeton. I responded,

"Oh, that phenomenon simply means that there are so many merchants who try to make money by selling such stuff. If you are in a country like the States, you would be smarter to invent some games and sell them to other people, instead of buying a game and playing it."

He sank into silence, and never spoke a word to me afterward. Possibly, he thought that I was chiding him for behaving like a child who was more interested in playing games than in studying. That was not my intention, of course. I was telling him jokingly an example of looking from a different perspective at an ordinary observation. But the key point is that he had no desire for learning something. He refused to be educated.

Here are two imaginary sequels to the conversation. One goes like this:

"That may be so, but it's enough for me to play a game."

"Did you find an interesting game worth playing?"

The other could be:

"Oh, I see. Does it apply to mathematics too?"

"Yes, of course. There are many people who are content to buy and to play cheap games of mathematics."

Whatever continuation he would have taken, we could have exchanged our ideas, but once he closed himself like a clam, nothing could be done. I didn't know how I could open it, nor have I found any way to do so.

As I already noted, the salary of a government employee was low. Senior members of my department had no ambition as research mathematicians, nor any jealousy towards younger people. They were kind enough to help us find ways to earn extra money.

One of them, Saburo Yamazaki, who had taught me Russian, arranged a teaching position for me at a cram school for those preparing to take college entrance examinations. In the Edo period there were many ronin, that is, samurai with no lord to serve. In the Showa period the students who graduated from high school but failed the entrance examination to college were called ronin. Almost all students at the cram school were ronin, though some of them had been accepted at less desirable colleges, and wanted to try their luck at the next year's examination in order to enter better schools.

At this school I was supposed to teach four 50-minute classes in the morning once a week, but I had to do far more than that because of the way the school operated, which was as follows. All such entrance examinations would be finished by the middle of March. Within two weeks a publishing company specializing in such a business would publish a huge volume containing all the problems with solutions. Each teacher of the school would get a portion of that volume, and choose the problems for teaching in the class. I don't remember how many weeks I taught, perhaps 13 weeks from April through July. I would choose 52 problems from what I got, and each student would be given a printed copy of the set of problems without solutions. Then I would teach to the students how to solve four problems in one class, and repeat the same four times.

The reader can imagine how nontrivial the task of choosing 52 problems in a short period would be, as I had to check the problems one by one. I am amazed by the fact that I was indeed doing that.

I discovered that many of the exam problems were artificial and required some clever tricks. I avoided such types, and chose more straightforward problems, which one could solve with standard techniques and basic knowledge. There is a competition called the Mathematical Olympic, in which a competitor is asked to solve some problems, which are difficult and of the type I avoided. Though such a competition may have its raison d'être, I think those younger people who are seriously interested in mathematics will lose nothing by ignoring it.

At the cram school there was a student who always came to me after class and asked some questions, which were of the type every teacher would have liked to avoid. Simply put, he was a nuisance. When the last term of the academic year at the school ended, I felt much relieved and was happy to have severed my connection with him. After spring break, I went to the classroom at the university on the first day of the new academic year, and to my unpleasant surprise, found him in the first row of the room. I told this to my colleagues, all of whom showed heartfelt sympathy at my misfortune. Conversely, he might have been unhappy with the fact that he would be taught by a cram school teacher even at university.

The cram school operated even in the summer, and attracted many students from various areas of the country outside of Tokyo. Many, perhaps most, of them were not ronin. For whatever reasons, the teachers were paid somewhat better in the summer session than normal. Aside from the salary from the cram school, I had a small additional income produced as royalties from a few textbooks that I co-authored with some of my colleagues, and I was able to save a sizable amount of money, which I thought would become necessary later. That indeed happened as I expected, as I narrate below.

As I wrote before, I corresponded with Weil in 1953. The International Symposium on Algebraic Number Theory, Tokyo–Nikko, was held in 1955, and Weil was one of the nine foreign participants. I naturally met him, and since then he recognized me as a promising young mathematician, far more so than the older people in Japan did. In early 1956 he encouraged me to come to Paris in 1957, as he would spend the academic year 1957–1958 there. Following his suggestion, Henri Cartan secured for me a position of chargé de recherches at the CNRS (National Center of Scientific Research). Until then many a Japanese mathematician had gone to France as a *boursier,* that is, a student studying with a scholarship, as well as travel expenses, provided by the French government. But that was not suitable for a researcher, and Weil conceived the idea of a position at the CNRS. In any case, I did not initiate the idea of going to France. In mathematics, I do what I like to do. I never received any direction from anybody. Otherwise, I often leave the matter to take its own course.

Though the French government would give me a salary as one of their employees, they would not pay travel expenses. At that time the air fare from Tokyo to Paris was almost 700 dollars, more than my yearly salary at the university. A senior member of my department advised me to apply to the Ministry of Education for funding, which I did, but after a long delay, it turned out that I would get nothing from them. Fortunately my savings were equal to that amount. However, with the help of another senior member, I applied for funds from the Asia Foundation, which was far quicker, and gave me 250 dollars. I remember that the official in the Foundation who handled the matter said that he had a brother who, as a student of A. A. Albert, worked on "higher abstract algebra."

The salary I received in France was more than decent. I even got an end-of-year allowance, which I used for the airfare from Paris to New York. After that I made another air travel from New York to Tokyo, which I paid for myself. The Japanese government gave me no travel expenses for these air travels, though I obtained something for my travel from Paris to Edinburgh, as I was officially a delegate from Japan to an international meeting. Interestingly, I was still being paid as their employee while I was in France and the States. This applied to everybody who was in the same situation as I was. They did not care about the fact that I was being paid by foreign institutions. Perhaps they thought they must keep paying the bread-winner on whose income his family depended. For some bizarre bureaucratic reason, they would have been happier had it been such foreign institutions, instead of me, that paid the travel expenses. In any case the time I spent in choosing cram school problems and in dealing with the nuisance student was not wasted.

Returning to the story of my going to France, the personal notes I made at that time include a sentence: "It is comical for me to be learning French and going to Paris." In other words, I wasn't expecting that I would become such a person. A senior professor, apparently sensing my impassivity, said, "You don't seem to be very happy about going abroad."

## 13. In Paris

I boarded a twin-engine Air France plane at Haneda Airport on November 23, 1957. There were fewer passengers in those days, and I was chauffeured by an Air France car from my home in Mitaka to the airport. In fact, there were a few Japanese passengers besides me, but I was the only one who went as far as Paris. The plane stopped at Manila, Saigon, Bangkok, Karachi, Teheran, and Rome. I visited Rome twice on later occasions, and even Teheran thirty-seven years later, but I never returned to the other cities after this journey, which took forty-eight hours.

It was midnight when I arrived at the Invalides air terminal in Paris. The winter there was depressingly dark. I had been taking lessons in French composition and conversation with a French woman in Tokyo. She told me, "Your heart may be leaping in expectation of being in Paris soon, but in reality, you will find yourself in utter darkness, which will make you weep, longing for Japan." Though my heart was not leaping, she was indeed correct as to the darkness of the city.

After ten months in Paris, I went to the United States, and in retrospect, I think it was to my advantage that France was the first foreign country in which I lived. From Paris I was able to travel to Switzerland, Germany, Italy, and Scotland, and had interesting experiences in those European countries, which I will describe later. I think this was probably better than going first to the United States, as it enabled me to have a wider perspective of life in general.

The subway system in Paris at that time was not much different from that of today. However, there were several things that we do not see now. For example, there was a heavy door at the end of the passageway to the platform, which would close at the arrival of the train, to prevent latecomers from rushing. The door of the train would close automatically, but would not open unless a passenger opened it by hand, which I found odd and inconvenient, as all such doors of Japanese trains opened and closed automatically.

The name of each subway station, as well as those of next ones, was clearly displayed in large letters, but that did not apply to the stations of ordinary inter-city railroads. In fact, there is a joke about a Japanese gentleman who went to France. After reaching Marseille by boat, he took a train to Paris. Looking out of the window, he made a great discovery: all French stations have the same name, *Sortie* (exit). I later found out that this was indeed the case, as the name of each station was almost impossible to notice compared with the far more conspicuous SORTIE.

I found one new thing that I did not have in Japan: ballpoint pens. Strange as it may sound, I had not used such a simple and practical writing instrument in Japan. Even now I am puzzled: Why was Japan so slow in introducing it?

At that time there were many alcoholic beggars on the streets and in subway passageways, holding out their trembling hands for money. A large warning was posted inside each subway train, that said, "Drinking a liter of wine a day will make you an alcoholic."

I once had lunch at a restaurant whose clients were mostly laborers working nearby. Many small bottles filled with red wine, without corks, as well as glasses, were placed on the table. Apparently those bottles were originally used for beer. A client would order something from the menu and would drink as much wine as he liked by pouring wine into the glass from the bottles. His bill would note the number of bottles he consumed. Interestingly I noticed escargots (snails) on the menu there.

Tomoo Tobari, one of my high school classmates, who later became the director of the Maison du Japon in Paris, was studying in Paris at that time, and kindly told me many French customs I didn't know. However, he was very impressed by my story of eating at such a place, as he never had such an experience.

About a month earlier, perhaps in March, after a seminar at the Poincaré Institute, Weil, Roger Godement, and I took a rest at a café on the Boulevard Saint Michel. The two French gentlemen ordered beer, but I had a vin chaud (warmed red wine), as it was a chilly day. I later told Tobari that Weil grimaced at me as if he was contemptuous of my choice. He said, "Well, a vin chaud may be called a widow's nightcap." He was good at making such interesting expressions.

I was glad to see Chevalley again, whom I had met in Tokyo in 1953. He was an assistant professor at Princeton in 1946 when he wrote the book, *Theory of Lie Groups,* that I mentioned before. He then became a professor at Columbia and moved back to Paris afterward. While in Princeton, he divorced his first wife. One day around that time a colleague of his asked him, "I didn't see you last week. Were you out of the town?" to which he replied, "I was in Reno." "Oh, did you give a talk there?" This used to be one of the standard Princeton jokes, but is now forgotten, I think. He was friendly with Iyanaga, and visited Japan a few more times; he also taught at Nagoya University. He liked to play the game of go, and I played against him at least twice, in Tokyo and Paris. At that time all Japanese mathematicians who played go, except me, were more advanced than Chevalley, which was why I became his opponent.

As for mathematics, I was working on a few projects in Tokyo before coming to Paris. One of them concerned the modular function field and modular correspondences. By July 1956 I was able to prove definite results, but it took several more months to formulate them in a well-organized theory. I eventually included it as Chapter 9 of the book in Japanese, which I was writing in collaboration with Yutaka Taniyama and which was published in July 1957. When I informed Weil of this theory, he advised me to make a summary and publish it in *Comptes* rendus, and so I sent him my manuscript sometime before my departure. It appeared around the same time as my arrival. The longer version, also in French, appeared in 1958. I eventually developed the ideas of this paper into a theory that formed an important part of my book which would be published in 1971.

The other was the right definition of the field of moduli of a polarized abelian variety, which was essential in the theory of complex multiplication. I found it around October, 1957, and it was the first topic I discussed with Weil at the Institute, and my mathematical life in Paris started smoothly, and I would think he was pleased. After this, I began my work by applying that idea to the easier cases of what I was thinking. The reader will find more about my relationship with Weil in Section A5.

I had a vague plan, but did not have the whole picture, though I thought I would be able to manage if I proceeded in the same direction. In other words, I did not know how to approach my problem in a precise way, but I instinctively knew that I was doing the right, if not the best, thing at each instance.

Speaking in general, there are many mathematical problems every mathematician in that field knows, and many researchers attempt to solve them. What I was doing was completely different from such an endeavour. I did not know in advance what kind of mathematical statements I was going to make. In each short time period I knew what I was able to do, but at the same time I felt that what I had done was merely a small portion of a much larger theory. Namely, I realized what I had wanted to do only when I finished it, which took almost ten years. Also I was the only person at that time who was interested in my subject.

What was that subject? I refer the reader to a short article on my work titled "Response," which was published in 1996, and is included in this book as Section A4. Here let me just note that my work is related to the Fuchsian group that Poincaré discovered in 1886 in a strangely sudden way, its generalization by Fricke, and to Hecke's dissertation of 1912. By March, 1958 I was able to handle the easiest case that concerned Poincaré's group. I chose the result as the topic of my talk at the International Congress of Mathematicians held in Edinburgh. I don't think there was anybody in the audience who was able to understand it.

As for the part related to Hecke's work, he had a strongly negative opinion on the kind of mathematics I eventually did, and therefore I wrote in that article of mine in 1996: "I can assure the reader that I had no intention of humiliating Hecke posthumously." This is of course a joke, but to confess, when I finished my work "Construction of class fields etc." (1967), I truly wished that Hecke had been alive then. But to my disappointment, Hecke had died twenty years earlier at the age of sixty.

Since I am noting only what I was able to do, the reader might have an impression that I did many things effortlessly, but of course that was not the case. No mathematician can and could do so. Every success is achieved only after many attempts of trial and error, and one is often rewarded by a failure.

To continue my narrative, let me first return to that period, which was indeed exciting. As to my French, Weil later said, "At first I thought you spoke French better than I had expected, but you never improved during your stay in Paris," which was true. I was studying French seriously before going, but after arriving there, I paid far more attention to my mathematical work than the language. He encouraged me to read French newspapers and see movies, and I followed his advice on the latter, but not on the former. There was a tabloid-sized publication that presented popular films in comic strip format, which I found at a barbershop. Perhaps I would have benefited from reading it, but I didn't do so. I also had the chance to socialize with a French girl, which could have made my French better, but somehow I did not wish to make my life too exciting, and did nothing in that respect. It is said that Hemingway hardly spoke French while in Paris, in spite of his novel, *A Moveable Feast.* Therefore my French may have been better than his, though that comforts me little.

## 14. In Princeton

In the fall of 1958, after ten months in Paris, I became a member at
the Institute for Advanced Study in Princeton, where Weil became
professor at the same time. So I saw him almost every day in the
next seven months. He had been investigating algebraic groups in
Paris, and he continued his work in Princeton, taking his aim at a
better formulation of Siegel's theory of quadratic forms, whereas
I was interested in modular forms. However, whenever I went to
him to talk about my work, he listened to me in earnest.

He was known for his impatience, and quite often shouted at
somebody. I think he shouted at me twice. I never shouted at
him, but once I told him what I wanted to say in a formal way,
which happened around 1970. I developed a certain theory, and
somebody reformulated it in his own fashion. Weil then said to me,
"I like his formulation, and in fact I prefer it to yours." It naturally
got on my nerves; besides, I did not think much of the formulation
in question, and so I said to him, "I find nothing praiseworthy
in his formulation; he and you are perfectly free to rearrange or
rewrite the work of somebody else. But I am not that type; I
simply try to discover new things and develop them into a new
theory." He was silent for a few minutes, and our conversation
turned to other subjects.

Speaking about my mathematical work in Princeton from 1958
to 1959, I continued to develop the ideas about the field of moduli
and other things that I explained in the previous section, but I
wrote up the results only after I returned to Tokyo. In addition to
these, I had a new idea about the periods of automorphic forms
of higher weights, and finished a paper on this topic in the spring
of 1958. I think Weil was impressed by this work, but I later
discovered that I had to view the phenomena discussed in the
paper from a different angle. Therefore it was not a complete
success, but after many years such periods would form one of the

important themes of my investigation, on which I would be more successful.

I insert here some interesting experiences I had in Princeton unrelated to mathematics. One day a woman working for the United Way (Red Cross) visited my apartment, and solicited contributions from me. As this was new to me, I asked her what the reasonal amount would be. She said, "Well, one dollar would be suitable for a member of your age." One dollar in those days was a fairly good sum, but I felt somehow that was a little miserly, and so I gave her two dollars. When I met my next door neighbor a few days later, he complained to me, "I just wanted to hand her one dollar, but she said that you gave her two dollars, and .... " I don't remember what the final outcome was.

Shortly after my arrival at Princeton, I checked in the Yellow Pages whether any French restaurant existed in the town. Having found one, I went there one evening. The menu looked French all right, and I ordered something. But then the waiter asked me if I wanted to drink coffee immediately, which was a total shock to me. In those days it was common, or rather standard, for Americans in a restaurant to drink coffee during dinner. As everybody knows, this has become far less common, but at that time I thought Americans had completely different kinds of taste buds.

I have to note one thing more important than coffee, still unrelated to mathematics. The one-room apartment I occupied was furnished, and had a kitchen with a small refrigerator, but the most essential feature was that it was well-heated and had a thermostat. I could also take a bath whenever I wanted. For someone like me who had been shivering all one's life, the place was a paradise. There were luxurious apartments in Tokyo, and in fact my teacher in French conversation lived in one of them, but such a place was far beyond the reach of college teachers.

There is a small lake called Carnegie Lake in Princeton. A brook flows from some point near the Institute to the lake. In winter both lake and brook are frozen, and one can skate on the lake. So I bought a pair of skates and enjoyed skating on the lake. One of the fellow members kindly taught me how to skate. We reached the lake by walking on the frozen brook. I especially liked the fact that I could immerse myself in warm water in the bathtub immediately after returning from the lake.

In addition to the heating and warm water there was one more thing which I viewed as an advantage of being in Princeton. Though at that time I was not related to Princeton University, I was able to obtain a membership of the University Store and to get discounts for my purchases there. One day in that store I bought a copy of the English version of *Mozart, His Life and Work* by Alfred Einstein, which is now considered a historically important literature on Mozart, and which I read avidly. The Japanese translation of this book was published only at the end of 1961.

I went back to Japan in 1959, and returned to Princeton in 1962. Since then I read many books in English which I would not have read had I stayed in Japan. Some of them were borrowed from the libraries of the town and the university, and some bought at the University store or other bookstores. I also found a second-hand bookstore nearby with a good selection of reasonably priced art books. There were of course several large bookstores in Tokyo where one could buy books in foreign languages, but the matter was much simpler in Princeton, as the books greeted my eyes without any special effort. In addition, Princeton, though a small town, has a theater which presents performances of world-class musicians and stage artists. Indeed, I later enjoyed the pantomime of Marcel Marceau several times there. Therefore I could lead a more interesting life in Princeton than in Tokyo or Osaka, at least in that respect.

Continuing my story at Princeton, I must note that I saw Robert Oppenheimer, who was the director of the Institute then. He is principally known as the person responsible for making the first atomic bomb. Let me present here what I remember or heard about him which cannot be found in his standard biographies. When I arrived at the Institute, I was asked to make an appointment to see him. It is my guess that he thought it was his duty to meet every new member. So, one day I entered his office and shook hands with him. Apparently he was checking my file, and said, "I gather that you were in Paris before coming here." We talked a little about some mathematicians both of us knew. I don't remember much beyond that. He was somewhat clumsy at such matters, for a man of age 55. I was 28 at that time.

I don't think I talked with him after that during my stay there. I saw him again some time in the summer of 1959, when he visited Tokyo. Shigeharu Matsumoto, the director of the International House in Tokyo, held a lunch party in his honor, and all previous Japanese members of the Institute who were in Tokyo, along with their spouses, were invited. I don't remember much about that party, except that Matsumoto was wearing a tweed jacket in spite of the warm weather. There was no awkward moment, as Matsumoto tried his best to make the party proceed smoothly, but nothing interesting was talked about. Oppenheimer was reticent, and did not volunteer to present any topic for conversation.

In the summer of 1963 I stayed with my family in Boulder, Colorado for two months. We rented the house of his brother, who was a professor of physics at the University of Colorado in that city. He looked very much like his elder brother; even the way he smoked a pipe was a faithful copy of that of his brother.

These are all what I know directly about Oppenheimer and his brother. Since then I heard many stories about Robert from various people. They were about his personal character, not his academic achievements. Clearly nobody liked him. According to what I heard from many sources, he was contemptuous of mathematicians, saying that "They are uncultured."

At the Institute, it was customary for a new member to give a one-hour talk on his work. It is said that Oppenheimer would call such a speaker into his office beforehand, and let him explain the contents of his intended talk. Then he would ask what the strong points and weak points of the work were. Since this was done by the director, the young scholar would dutifully comply with his request.

On the day of the lecture, he would take a seat in the first row in the lecture hall. At the end of the lecture he would turn to the audience, and would comment that the strong and weak points of the work were such and such, exactly as the speaker had explained to him. This may not be factual in its totality, but I believe it contains more than a grain of truth.

A well known math-physicist Eugene Wigner was in our department, and so I occasionally talked with him. He was pompous and took himself very seriously. That is the impression shared by all those who talked with him. At a departmental party, he asked me what kind of mathematics I was doing. He asked that question as if he met me for the first time. At that time I had been a full professor and his colleague at least for six years, perhaps more. I told him vaguely, "Well, mainly things related to modular forms." Then he said, "Oh, modular forms; we physicists don't need such" in a very contemptuous tone.

I can add that there are some physicists who are interested in modular forms. Once Edward Witten attended my graduate course on Siegel modular forms, and often asked sensible questions in class.

Of course I met many pleasant people. One of them was Lyman Spitzer, a famous astrophysicist, whose department building is just a few hundred feet north of the mathematics department building, which includes a tower whose total number of floors is 16. He used to climb mountains when he was young. After his retirement, he would come to the math building and walk up the stairs of the tower from the street level to the twelfth floor. I had (and still have) my office on the fifth floor, and used to walk up the same stairs, without using the elevators. Therefore I often met him on the stairs, and had casual conversations with him; he would tell me that he was just back from Japan where he spent the previous week. And we would talk about some mutual acquaintances. There was nothing serious, but I could easily tell that he was a gentleman, and that was the unanimous opinion held by all those who knew him.

Let me now describe my impressions of Teiji Takagi, who played a major role in the development of class field theory. He had been worshipped as the very greatest mathematician in Japan. There was a dinner party of Japanese mathematicians in 1955. As mentioned in Section 12, several foreign mathematicians visited Japan that year, but they were not there. He and I were placed at different tables, but I was able to hear what he was talking about. He was amusing himself by telling a stale joke, and nothing interesting was uttered from his mouth. Shortly after that, four or five young mathematicians of my generation, including myself, visited him at his house to talk with him. He was eighty and hard of hearing, and so whenever one of us spoke, a middle-aged woman who was his relative restated the content in a tone and pitch he could understand.

He was anything but a conversationalist. He didn't show any warm feeling to us. I searched for some topic, and recalled that he had said, around 1940, that mathematics of the time was in a transitional period. So I asked him what he meant by those words. I did so merely because I could not find anything better to ask. When this was told to him by the woman, he was clearly upset. He said something fast in an angry tone. I don't remember what exactly he said; perhaps he refused to explain.

I heard him speak only on those two occasions, but I was completely disappointed. Confucius divided men into two categories: *kunshi* (a virtuous man) and *shojin* (a worthless man). There is a passage in *The Analects of Confucius:* "Kunshi is peaceful and not arrogant, whereas shojin is arrogant and not peaceful." I thought I had seen an example of someone who was arrogant and not peaceful. Later I even heard that he was disliked by his close relatives.

There is an old Chinese saying: "If the observations of ten people are the same, that should be right." It should be noted however that this refers to one's personal character or moral principles, not to the artistic or scientific value of one's work. I knew Artin, Chevalley, Eichler, as kunshi; Siegel was mean, but not sho-jin. Richard Brauer was neurotic, and I cannot include him in the category of kunshi, an opinion held also by Iwasawa. I note here Tadao Tannaka as kunshi.

## 15. Return to Tokyo

I returned from Princeton to Tokyo in the spring of 1959. I was driven from Haneda Airport to my old home in Mitaka, and was depressed by the poorness of the houses lining the streets seen from the car window, which I had not noticed before. Japan was still a poor country; I was poor too, which would become a major problem in the next few years of my life, but I first have to note one fact: I married Chikako Ishiguro in that summer. I had known her since 1953, but I was not seriously thinking of marrying her; I merely had a vague idea that it might happen that she would become my wife. Eventually it did happen to be so. I think that is an accurate description. Her ancestors served the lord of one of the poorest fiefs of Japan. The genealogical table in her brother's possession is so extensive that they might have hired a professional to produce it, a common practice in the Edo period. The most noticeable item on it is that one of them helped a fellow samurai avenge his father's death, which may or may not be fabricated.

She was born in Tokyo and raised in a house that was a ten-minute walk from my Ohkubo house. The school she attended was called Toyama elementary school, which adjoined Toyama-ga-hara. Therefore I married a girl from the same Kiri-ezu, but not of the same fief. There was a rivalry between her elementary school and mine, and the pupils of one school always spoke of those of the other contemptuously, a phenomenon common to any pair of nearby elementary schools in Tokyo. However, nobody compared us with Romeo and Juliet.

I naturally resumed my teaching at the University of Tokyo, but that became a pretty heavy strain on my life, which eventually made me leave Tokyo. Before coming to that subject, I now describe my mathematical work in the two years after my returning to Tokyo. In July 1957 I had published a book *Kindai-teki Seisu-ron* in Japanese (Modern Number Theory) written in collaboration with Yutaka Taniyama. I wrote a short preface of the book whose first half may be translated as follows:

"The progress of algebraic geometry has had a strong influence on number theory. It has been an important problem to establish a higher-dimensional generalization of the classical theory of complex multiplication by Kronecker and to complete the work left by Hecke. By means of the language of algebraic geometry we can now add new knowledge in that direction. We find it difficult to claim that the theory is presented in a completely satisfactory form. In any case, it may be said, we are allowed in the course of progress to climb to a certain height in order to look back at our tracks and then to take a view of our destination."

Simply put, the book gave the theory of complex multiplication of abelian varieties to the extent we could achieve with what we knew at that time. In doing so, the theory of reduction modulo $p$ in my paper I sent to Weil in 1953 played an essential role. In fact, I developed the theory with the intention of applying it to complex multiplication. As I said in the preface, there were several unsatisfactory points in the book. One of them was the proper definition of "the field of moduli," which I discovered only in October 1958 and which I told Weil immediately after my arrival in Paris. Thus, the first thing I did after coming back to Tokyo in the spring of 1959 was to write the whole theory in English in a better form by using this new definition.

We had actually planned an English version, but nothing was done except for a short section I wrote in English on differential forms on abelian varieties. Sometime in 1957 I handed it to Taniyama, who died in November 1958. It was returned to me when I met one of his brothers. I eventually published the book in English as a collaborative work with him in 1961, but actually I wrote everything alone, and he was not responsible for the exposition.

I had known that he was not a careful type, but after starting this project in 1959 I realized that the problem was more serious than I had thought. Indeed, I had to throw away many things he wrote in that book in Japanese. In my article about his life published in *Bulletin of the London Mathematical Society* (1989), I wrote: "Though he was by no means a sloppy type, he was gifted with the special capability of making many mistakes, mostly in the right direction." I also wrote in the preface of the 1961 book in English: "The present volume is not a mere translation, however; we have written afresh from beginning to end, revising at many points, and adding new results such as §17 and several proofs of propositions which were previously omitted."

Thirty-five years later in 1996 I published a book, of which I was the sole author, the first half of which was a revision of this book, and the last half of which contained new results on the periods of abelian integrals. Although this subject is related in various ways to other topics I investigated later, I do not talk about them here.

Aside from the 1961 book, I published a joint paper with Shoji Koizumi toward the end of 1959. Some time in the spring of 1960, I received a letter from Alexander Grothendieck, whom I had met in Paris in 1958. In the letter he said that one of our propositions had a counterexample. We realized that we had forgotten to state a condition, and I wrote him that the condition should be added, which settled the matter. (See Notes to [59d] in my *Collected Papers*.) In fact, in an earlier letter to me he expressed his skepticism about a result in my 1955 paper on reduction modulo $p$. In my reply I explained that there was a certain strong condition which was stated but which he might have overlooked. I think he was convinced. So he was one of the few people who really read my earlier papers. I also remember him as a man with a completely shaved head, who looked very much like the actor Yul Brynner. According to some French mathematicians at that time, he got the idea of shaving his head from the actor, but I am not certain. I saw him again at the International Summer School held in Antwerp in 1972, where he was playing an anti-NATO game; it was so childish and silly that I will not write about it.

In Tokyo from 1959 to 1960 I was preparing the above 1961 book, which I enjoyed, but I began to detest teaching at the University of Tokyo, as I said before. Not that I did not like to teach in general; only, the system was so irrational and stupid that both students and teachers were unhappy. After finishing the manuscript of the book on complex multiplication of abelian varieties, whose preface is dated February 1960, I wrote a paper that gave the full details of the results concerning the algebraic curves associated with the Fuchsian group of Poincaré, on which I talked at the Edinburgh Congress in 1958.

The Japan–U.S. Security Treaty, that had been signed in 1952, was revised in June 1960. In that month there were large-scale demonstrations by labor unions and students who were against the treaty. As a result, the university was practically closed, but without going into detail, I merely point out that the anti-U.S. sentiment among Japanese citizens had been intensifying in the ten years prior to 1960, because of various events, most of which seem to be forgotten now.

I clearly remember one of them that happened on January 30, 1957. The American Army stationed in Japan had several training grounds, one of which was in Gunma Prefecture. Peasants used to enter the grounds, gather the spent shell casings, and sell them to scrap metal dealers. On that day, an American soldier shot one of these peasants to death from a thirty-foot distance, for no reason at all. He was caught, tried at a district court in that Prefecture in August, and received a three-year prison sentence suspended for four years. But he was able to return to the States in that December. Evidently, the Japanese government pressured both judges and prosecutors. There were several more incidents of a similar nature. In fact, there were 9998 officially reported cases of damages, injuries, and homicides caused by the American Army by 1958.

I did not join the demonstrations, but one day I discovered in the newspaper that one of the arrested demonstrators was a student in a class under my supervision. So I visited the Metropolitan Police Headquarters, and talked with one of the prosecutors to inquire about his status. Our conversation was neither hostile nor very friendly; I think the student was released a few days later.

I also note here an event that was related to my work. Some time in April or May that year, the student newspaper asked me to write something about what I was doing at that time. I agreed, and sent them a short article in which I quoted the passage of Poincaré's *Science et Méthode* that described his discovery of arithmetic Fuchsian groups, and said that my work was closely connected with his investigations done more than seventy years ago. But the newspaper, preoccupied with reporting the rapidly changing political situation, had no interest in my article, and as a consequence I never heard any word from them afterward. Also, I think Poincaré meant nothing to them, though they certainly knew who Karl Marx was.

Aside from my own work, I organized a seminar on the arithmetic of algebras. (Here "algebra" is a standard, if odd, technical term, which was once called "system of hypercomplex numbers.") It lasted from October 1959 through December 1960. There were about eight speakers that included Michio Kuga, Hideo Shimizu, and myself. Eventually, a volume of mimeographed notes was published in early 1963 by the Mathematics Department of the

University of Tokyo. I wrote its preface, the last passage of which reads like this.

"Needless to say, the problems treated here naturally continue to the arithmetic of algebraic groups, which has just begun to be developed. Though our principal motive for this seminar can be explained in that fashion, it is unnecessary to put much stress on this point. I believe that this old subject which had been neglected for some time will be able to provide the reader who has no preconception with some new ideas for mathematics of tomorrow."

The seminar was a success to the extent that it presented some standard facts with which researchers must arm themselves before entering a new frontier of mathematics. While in Paris I had noticed a 1938 paper by Eichler concerning what we now call strong approximation, whose importance was not recognized in Japan, nor in the States. This was one of the "standard facts" that I wanted to include, but time constraints prevented us from doing so. I also thought that I had to take the initiative in such a matter, as "the old men these days" could not be counted on.

More importantly to me, in the fall of 1960, I got a crucial idea, by means of which I could handle the curves associated with Fricke's generalization of Poincaré's group. Though I didn't know it at that time, I would eventually spend the next ten years of my mathematical life cultivating this idea into a major theory.

I was also developing the theory of Hecke operators for the symplectic group of degree 2. Among the main results were that there is an Euler product of degree 4, and that the product has a congruence relation. In January 1961, I gave a series of lectures on these at the Mathematics Department of the College of General Education, the University of Tokyo. I published a short summary in 1963. A mathematician wrote a paper in which he claimed to have discovered the Euler product independently. In fact, he was among the audience of my lectures and told me afterward that he obtained a product of degree 5. I don't know when he was able to correct that wrong result.

## 16. One Year in Osaka

In the spring of 1961 I moved to Osaka University, induced by
Yozo Matsushima who was a professor there. Originally, I was not
so enthusiastic about going, but I had been tired of dealing with
the bureaucracy in Tokyo, and so I decided to go, expecting that
at least something new would happen. Matsushima, who was nine
years older than I, and his wife Fumiko became our intimate family
friends. I played the role of his confidant, as he freely expressed
his unflattering opinions on the professors at Nagoya and Osaka.
I also wrote a joint paper with him, which was published in 1963.

At that time there were of course mathematicians in Japan
who were doing serious mathematics, but their number was small.
Excluding those few, the level of professors and students was incredibly low. There was no good academic environment either.
Speaking especially about Osaka, immediately after arriving there,
I keenly felt the cultural backwardness of the city. I had been typing my manuscript on my typewriter, and naturally I needed new
typewriter ribbons from time to time. In Tokyo I had no problem,
as almost every stationery store of medium size had them. But in
Osaka I was unable to find a shop selling ribbons. There were many
trade companies and colleges in Osaka, and they of course used
typewriters. I guess they purchased all stationery items wholesale,
and practically no individual bought typewriter ribbons. Besides,
there were relatively few bookstores and stationery stores compared with Tokyo; instead there were many shops for electrical
appliances. The whole atmosphere made me think that I had
come to an intellectual wilderness. There was one more strange
fact. I was an associate professor at Tokyo, and became a full
professor at Osaka, but the government still paid me as an associate professor according to their rule. This surprised me as well
as everybody with whom I talked. In fact, when I became involved in preparing the entrance examination problems at Tokyo,
I quit the cram school. I taught part time at a women's college for

the academic year 1960–1961, but after coming to Osaka, my only income was from Osaka University, which was roughly 40,000 yen per month, barely sufficient for a family of three; my daughter had been born in May 1960. Three years earlier in 1958 my monthly salary at CNRS was 90,000 francs, perhaps equivalent to 70,000 yen, possibly more.

Anyway, while in Osaka, I began to think that I must go to the United States. When Weil visited Japan in the spring of 1961, I asked him to find a position for me in the States, which was the only occasion I sought a job in my whole life. As I already wrote, the apartment I had in Princeton was well heated, which was bliss for someone like me who had been shivering until then. I had known that Kyoto was much colder in winter than Tokyo. Osaka was not as cold as Kyoto, but it was quite chilly to be in a Japanese house that was not well insulated. Once I was working at a desk with an oil stove at my side, and I put my legs so close to the stove that one of my socks was burned.

Many years later I was asked by some Japanese people why I decided to live in the States, and I always answered, "Because it was cold in Japan." Almost all questioners took this as a metaphor. Though that was partly true, it was literally cold in Japan, at least for me. Since 1956, I had a certain mathematical vision, and that became clearer by the time I went to Osaka. Also, I had a definite idea about Fricke's group, and I might have been able to complete the work by staying in Japan, but I did not like the prospect of doing so by burning my socks.

Weil, having had gone through difficult times himself, knew how to handle such matters; also he was widely recognized. Eventually he succeeded in persuading the Princeton mathematics department to offer me a professorship. At that time the department had good people in topology and analysis, but not in algebra or number theory. I was familiar with number theory, algebraic geometry, and modular forms, and there was no such mathematician in the United States.

Not that I decided in 1962 to stay in the States permanently. Academically I had confidence in myself, but I was not completely sure that my family would be able to lead a happy life in a foreign country, though I was optimistic. One thing was certain from my viewpoint: I had something to do in mathematics, and I viewed the United States as the best place for achieving my aim. I used to tell my American friends jokingly that I had been exiled by the Japanese government, and was waiting for a letter of pardon. Of course they were formed by the type of people who would never have the idea of issuing such a letter, and besides, life as an exile was reasonably comfortable, and the keepers for the most part were friendly and warm.

Before narrating how I behaved as an exile, let me return to the point of my coming to Osaka. The western part of Japan that includes Osaka, Kyoto, Nara, and Kobe is called Kansai, which is culturally very different from Kanto, the eastern area including Tokyo and Yokohama. I had been in Kansai many times before 1961, but Chikako had been there only once.

There used to be a program for sixth-graders in Tokyo to make a one-week trip to Kansai. I don't know when it started, but it was an important yearly event for the pupils and teachers involved, held until around 1938 or 1939, when various difficulties caused by the war made such trips impractical. For this reason I went to Kansai for the first time only in 1949, when I was a university student. I remember one thing that impressed me: the color of the soil in Kyoto was yellow. The soil in Tokyo is black, but that is so only to the depth of one foot, and below that black soil is red-brown colored clay. This fact was firmly registered in my mind, as I had dug many shelters during the war.

As for Chikako, when she accompanied me to Osaka in 1961, it was her second time in Kansai. We lived in an old house owned by the university. It was in Ishibashi, one of the satellite cities of Osaka. The rent was low, but the house was full of cockroaches, which I had never seen in Tokyo. Our acquaintances, having heard the name Ishibashi, would say, "Oh, you live in the most desirable residential area," which was true, but I was unable to find a proper response, as my place was far less than desirable.

In Osaka, it was customary for a housewife to buy fish in the morning, whereas in Tokyo, it was in the late afternoon to do so. Therefore Chikako was puzzled by the fact that there was practically no decent fish to buy in the shops, until Fumiko explained that custom. Chikako could have said, "Today, I'm afraid I am short of money," as the second wife of my great-grandfather did. However, she had no silk robe to sell, and so she was not in the position to buy a large fresh bonito. As for myself, I intended to have another master, and therefore I was definitely not a loyal samurai.

There are a few notable mathematicians who were imprisoned. Hans Maass, who is known for his theory of something called Maass forms, is one. He was friendly with me, and told me that he had served a prison term just before the war's end for a political cause. That type of experience makes one's biography more interesting. To tell the truth, I once stayed in a prison for a short period. That happened some time in January or February of 1962. I was one of the professors responsible for preparing the entrance examination problems at Osaka University. For security reasons, such problems were printed in a large prison situated south of Osaka. So, one day, a few professors including myself went there to proofread. We arrived at the prison enclosed by high walls, passed through several heavily guarded gates, were given special passes, and eventually reached a large hall, where prisoners were operating printing presses. We read the proofs in that hall surrounded by criminals, who, we were told, were not of a dangerous type. I think we had lunch in another room waiting for the final proofs. We spent more time waiting than proofreading, and I think we were in that prison about two and half hours. This experience left me with a strong impression, and at least enabled me to join that small group of mathematicians, who had eaten in prison.

Our life in Ishibashi lasted only one year. In those days it was not easy and was certainly costly to travel between Japan and the States, and I did not expect to come back to Japan soon. So, one fine day in May, 1962, our family of three went to Kyoto for sightseeing. Avoiding the places that attracted many tourists, we visited Koryuji, Daikakuji, Tenryuji, and Kokedera, all old Buddhist temples known for their beautiful gardens, and also went to Arashiyama. I had the satisfaction of making a memorable excursion with no sentimental feeling that we might have been in those places for the last time.

## 17. At Princeton University

In September that year we came to Princeton, the town I left three years earlier. The streets looked the same as before, but this time I was associated with the university, where I began teaching. The chairman of the mathematics department was Albert W. Tucker and John C. Moore was co-chairman. After a few years, John Milnor became the chairman, but I had no problem with anybody. After teaching one term, I had the impression that I was not disliked by the students. Every senior at Princeton was (and still is) supposed to write an expository paper, called a senior thesis. There were a few students who chose me as the advisor of their senior theses, and it was so in almost every succeeding year in which I taught upper-class courses. Since I had a stock of easy but interesting topics, I welcomed them. Besides, I liked to talk with students face to face, more so than lecturing. Thus my life as a teacher at Princeton was certainly more enjoyable than in Tokyo and Osaka.

After one semester of teaching, I naturally had to grade final examination papers of my course. Though I had enough experiences in Japan, I went to the department office to ask a secretary to show me a sample grading sheet. She was Ginny Nonziato, who after some years became the department manager, the person who supervised all secretarial matters. She had begun to work in the department at the age of eighteen, and the typing of Siegel's *Lectures on Transcendental Numbers* (Annals of Mathematics Studies No. 16) of 1949 was her first substantial job. She kept a copy of that book as a memento on her desk and showed it to me.

Her sample grading sheet was Tucker's, which was ordinary, and I graded the papers as I had been doing in Japan. One day two or three years later, she called me into her office and showed me the grading sheet of a newly appointed instructor. More than half of the students in his course were given failing grades, which she considered excessive. She therefore asked me to mend matters. So I went to him and explained the common sense of grading to him. He stubbornly resisted my persuasion, but I eventually succeeded in making his grades more reasonable. I later discovered that there were certain types of people who lacked common sense and consistently behaved against common sense. This instructor was one of them.

Ginny was clear-headed, and whatever I said, she understood it immediately and never asked me to repeat. She was kind and liked by everybody. In contrast to her, Agnes Henry, the department manager at that time, was very different. When I joined the department, I shared an office with an assistant professor. A desk intended for me had been brought in, and a desk lamp was on it, but the cord of the lamp was not long enough. So I went to Agnes and told her that an extension cord was necessary. She said, "Ah, you can buy it at the University Store," which I found strange. After my staring her in her face without words for half a minute or so, she said, "All right, I will get it for you."

At a dinner party some years later, I told this to a mathematician of my generation who obtained his doctoral degree at Princeton. He laughed and said, "That is typical of Agnes." She was known for her meanness. Incidentally, the person who shared the office with me was quite odd. He was indifferent and seemed as if he did not want to have any relationship with me. I don't remember any meaningful conversation with him. I later found that this impression was shared by many people who knew him.

In the summer of 1963 I spent about two months in Boulder, Colorado. A conference on number theory was held in August at the University of Colorado, and I was asked to attend. Also, it was kindly suggested that I arrive before the conference and give a few lectures. The matter was arranged by Sarvadaman Chowla who was at the university and Atle Selberg at the Institute for Advanced Study. I accepted the invitation and went there with my family. We had a wonderful and unforgettable time as I describe in a later section.

Among the participants of the conference were Bryan Birch, Hermut Hasse, Martin Kneser, Marc Krasner, Louis J. Mordell, and Koichi Yamamoto. I had met Krasner in Paris and Kneser in Edinburgh both in 1958, but I met the others for the first time. I became friendly with Chowla, and saw him on many later occasions. Mordell, at the age of 75, was the oldest. Clearly he took himself seriously. He would sit in the first row of the lecture hall, and after almost every talk he would stand up, face the audience, and give his reminiscences on the subject of the talk, all of which were unmemorable. He didn't do so after my talk, and I don't think I ever spoke with him except that we must have shaken hands. I had interesting mathematical conversations with Birch, which I explain in my second letter to Richard Taylor in Section A3. When I met Hasse, he expressed some words of congratulations, which somewhat surprised me. I never figured out for what work of mine he was congratulating me.

I continued my work in Colorado, and found a crucial idea about the problem that I had been investigating. I explain the technical aspect of this idea in Section A4.

As I already mentioned, I participated in a conference in 1964, which was officially called the Summer Research Institute on Algebraic Geometry, held at Woods Hole, Massachusetts, where I had my near-drowning experience as I narrated in Section 4. Several months earlier I had formulated a possible new trace formula which would generalize the Lefschetz fixed point formula. During the conference I told this first to John Tate, and then to Michael Atiyah and Raoul Bott. The latter two were extremely excited about my conjectural formula, which was completely new to them. Eventually they proved the case that concerned a map, whereas I formulated the formula more generally for a correspondence. At first they acknowledged that the idea was due to me. But interestingly they gradually tried to minimize my contribution. In fact, in 2001 Bott claimed that I was not involved in the matter, and later was forced to concede that he was wrong. However, Atiyah noted in one of the volumes of his complete works that they learned it from me.

At the conference, I gave a series of lectures, which included an introduction to the theory of automorphic forms and Hecke operators, which was new to almost all participants. Mimeographed notes of the lectures were distributed during the conference, and many years later printed as article [64e] in my *Collected Papers*. Jean-Pierre Serre, whom I had met in Tokyo and Paris, was among the audience, and kept asking questions on the most trivial points, which naturally annoyed me. Although I did not know it at that time, several French friends of mine told me later that he did the same to every speaker. However, he had behaved reasonably with me in Tokyo and Paris, and that changed drastically in 1964. Somebody told me that he had become frustrated and even sour. Much later I formed an opinion that he had been frustrated and sour for most of his life. As described in my letter to Freydoon Shahidi, included as Section A2 in this book, he once tried to humiliate me, and as a result gave me the chance to state my conjecture about rational elliptic curves. I now believe that his "attack" on me was caused by his jealousy towards my supposed "success" — my conjectural formula and lectures — at Woods Hole.

It is a generally accepted fact that Japanese people look younger than they really are. I am no exception. Let me narrate one of my experiences in that respect. When I came to Princeton in 1958, the mathematics department was housed in a three-story building called Fine Hall, and it was so until 1969, when the department moved to the present building, which is now called Fine Hall; the previous Fine Hall is now called Jones Hall. The first and second floors of the old Fine Hall consisted of the offices for faculty members and secretaries, and the top floor was the library. There was a building called Palmer, occupied by the physics department, which was connected to the second floor of Fine Hall by an overhead passage. Most mathematics courses were taught in classrooms in both buildings.

One day in the fall of 1963, I was entering Palmer through the main doorway. There used to be a guard sitting in a corner office in the entrance hall. He stopped me and said, "Don't you know that no freshman is allowed to enter this building?" Apparently he was teasing me, as I looked very young. "Is that so? But this is my second year here," said I.

I had a memorable experience of a different nature, which occurred a few years before my retirement at the age of 69. In the new Fine Hall there is a sizable hall in front of the entrance to the library. Some desks are set in the hall, and students can work on computers placed on the desks. The hall also functions as a corridor. One late afternoon in the early fall, I was walking there on the way home, when a student hurriedly stopped me and said, "I'm a freshman, and I have difficulties with my physics textbook; I wonder if you can help me." Somewhat surprised, I said, "Maybe I can, but do you know me?" "No, but I have been frustrated. When I raised my head, my eyes caught sight of you, and you look like a person capable of helping me," said he. I could not but laugh, and said, "All right, tell me what your problem is." He showed me the passage in question from the textbook. It merely concerned the problem of how to interpret a symbol of partial differentiation. I explained the point by using the blackboard on the wall of the hall, and that was that. I don't know whether he was convinced that he chose the right college for him. In any case, this time I looked like an old enough person capable of explaining mathematics.

Speaking about my own mathematical problems, I was working according to the plan I had had for some time. Whenever I published a paper in the period 1963–1965, the results were reasonably good, but they were not decisive. I had a definite technical idea only in the fall of 1965, three years after coming to Princeton. Using that idea, I finished writing a 102-page paper in June 1966, which I dedicated to Weil on the occasion of his sixtieth birthday. At the end of the paper there is a note "Received June 6, 1966." I believe that it was done deliberately by Fanny Rosenblum, the technical editor of the *Annals of Mathematics* at that time. She and her husband, Charles, were our good family friends. I was a guest at the party they gave in honor of Don Spencer's sixtieth birthday in 1972. While I was talking with Fanny, she smiled and whispered in my ear, "What's so great in becoming sixty," and we chuckled. I suppose she was much older than Spencer. Once, referring to one of the editors of the *Annals* at that time, she said to me, "It is really difficult to understand him. He is an inscrutable Anglo-Saxon." She was one of the most interesting women I met in my life.

The reader will find in Section A5 several episodes about Weil and his relationship with me. Let me insert here another concerning wine. I think it was in the mid-1970s. André and his wife Eveline were dinner guests at my place. Seeing the label on the wine bottle, she exclaimed, in an impressed tone, "Oh, Meursault!" In those days Meursault was already not for everyday use, but not as expensive as it is today. About a year later Chikako and I were their dinner guests at the dining hall of the Institute for Advanced Study. Sigurdur Helgason and his wife were the other guests. André looked at the wine list for a few minutes, but somehow he was unable to decide. Then Eveline said, "Why don't you let Goro choose?" Strange as it may sound, he passed the wine list to me, and I was happy to select two bottles. One of them was Hermitage blanc, but I don't remember the other. The main course was fish. It is quite possible that the bottle of Meursault helped her form an opinion that I was better than he in choosing wine.

I add here one more story about our dinner with Weil in the spring of 1987. Chikako and I were staying in Paris for two months that year; Eveline had died about a year earlier. As I had known the date on which he would arrive in Paris, I phoned him and suggested that we have dinner together some time. Until then, whenever we were in Paris, we were always treated as the Weils' dinner guests. In fact, in 1978, we together with our son and daughter had dinner in their apartment. Once I said to him, "This time you will certainly be my guest," to which he replied, "No, no, as long as you are in Paris, you will be my guest."

However, in his answer to my phone call, he said, "That's a good idea. Where would you suggest?" I cannot say this surprised me, as I had the impression that he had changed greatly after his wife's death, and I somehow expected that kind of reaction. Also I was prepared to suggest the place; eventually we had dinner together one evening at a Chinese restaurant on rue Gay Lussac, which was ten-minute walk from his residence. Although he certainly enjoyed our company, and he did so on several more later occasions, the change he showed made me feel that he lost a certain strength in his personality, and from then on I would be dealing with a different man.

As to the chronology of how I developed my theory in the 1950s and 1960s, the reader is referred to my article "Response" of 1996, which is reprinted as Section A4 of this volume. In it I mentioned my paper dedicated to Siegel on the occasion of his seventieth birthday. In this connection I quote here a letter from Eichler.

Basel, Jan. 1st, 1966

Dear Shimura,

I want to send my best wishes for this New Year, and those of my wife to you and Mrs. Shimura. We both thank you for your beautiful card. As Eckermann told me some months ago you are planning to come to this country in summer 1967. I am looking forward to seeing you again.

In summer and autumn I almost struggled to find something to write for Siegel's 70th birthday-volume. He would not enjoy anything. This kept me away from my plans on the $n$-dimensional trace formula, your formula! Only during these short vacations during Christmas I could work on this. There is much preliminary work to be done. Now I hope most of this is done, and I even have a definite conjecture on the way leading to the solution.

With all best wishes
Yours faithfully,
M. Eichler

Contrary to his opinion that "He would not enjoy anything," Siegel was actually able to appreciate my dedication, as he expressed his thanks in his letter to me dated May 15, 1967, which is reproduced in Section A4. In that article I also said that he was skeptical about the result I had in 1958. However, in his paper of 1968 (*Gesammelte Abhandlungen,* vol. IV, No. 86), he cited four papers of mine, and spoke of the deep results of Shimura, which may be considered his formal acknowledgment of my results. In "Response" I mentioned his persecution complex that he was not sufficiently appreciated by the younger generation. I had no such complex, and was not as unhappy as Siegel might have been. But I always thought that few people really understood my work. I knew that Chevalley, Eichler, Siegel, and Weil understood my work, and that was enough for me. Even so, on some occasions, I felt I was being unfairly ignored.

Let me now give an example. My family of three stayed in Zürich for about one month since late June in 1967. There was a well-known institute ETH (Eidgenössische Technische Hochschule) in that city, and Eichler kindly arranged a research position there for me. We lived in an apartment in the western suburbs. As Switzerland is a small country, we were able to visit almost every tourist site by an overnight trip. In those days it was easy to find suitable accommodations by asking for information at the tourist office of the railroad station of our destination. So we enjoyed almost every weekend with such a trip.

Chandrasekharan and Eckmann were the professors of mathematics at the institute, the former taking care of my stay there. Evidently he took himself seriously, but understood nothing but the type of analytic number theory on which he worked. He received me indifferently, and was not interested in getting to know me, or in what I was doing. I did not realize it at first, but as time passed, it became clear that he was treating me like a novice who had just published his doctoral thesis. The two papers I dedicated to Siegel and Weil were certainly beyond his comprehension.

While in Zürich I was able to continue my work as usual; we also went to Basel to see Eichler; I visited Karlsruhe and met Kubota and Leopoldt. Generally speaking, I enjoyed my stay, and I certainly made my wife and daughter happy, but at some point I became tired of being a second-class citizen in the mathematical community there. So I returned to Princeton a few weeks earlier than originally planned. I said nothing to Chandrasekharan at that time, but apparently he realized, probably much later, how I felt. In fact, I met him again at the International Congress of Mathematicians at Helsinki in 1978, when he said something to the effect that he regretted that he was unable to treat me in a proper way in 1967. If I were mean-spirited, I would have said, "I am in complete agreement with you," but I am the type of person to whom such a line occurs a few days later.

It may be necessary to explain why my work was understood by those four mathematicians I mentioned above, but not by the majority of other people. There are mathematical problems that most, or at least many, mathematicians know, and the solution of any such problem is praised as an important work. The primary reason for this is the prejudice that "The more difficult, the better." Suppose a certain problem has been unsolved for a long time, and somebody solves it. Then the solver is praised as a wonderful mathematician. That is so even if both problem and solution are not important. Everybody can appreciate the fact that a longstanding difficult probelm was solved, but nobody really cares about the true significance of the solution in the development of mathematics.

What I had done in the 1960s was completely different. My work that might have "humiliated Hecke posthumously" was not anticipated by anybody, even by myself. I knew that there would be something interesting, but I was able to formulate the results only when I proved them. Once stated, they were very natural, but to understand and appreciate them, one had to have sophisticated knowledge and also a good historical perspective. Around 1980 I told a mathematician much younger than I, "No mathematicians with American citizenship understood my work in the 1960s." He, who was an American, said, "That must have been true, indeed."

In spite of the fact that my mathematical work was little understood by the general mathematical public, I was often the target of jealousy by other mathematicians, which I found strange. I can narrate many stories about this in detail, but that would be unpleasant and unnecessary, and so I mention only one interesting case. Before doing it, let me first state here some of my observations in general. The person who is jealous has his mind full of irrational envy. His target does not know it, but becomes aware of it when one day an unpleasant event occurs. In almost all cases my action is not the cause; the person simply resents my existence. In a few cases in which my action is the cause, it is unrelated to the jealous person. Serre's attack mentioned earlier is a good example. Some of those who are jealous of me work in fields unrelated to mine. Many of them are very competitive; some take themselves seriously; some are frustrated; some understand my work, but some don't. There are many cases, but one thing is common: it is unpleasant, but I can do nothing about it. One cannot call it a negligible matter, as it becomes difficult for me to keep a normal relationship with the person. To do so requires considerable effort on my part.

The "interesting case" happened in the fall of 1979. In that year Apéry published a paper on the irrationality of $\zeta(3)$. There was a party at the beginning of the academic year given as usual by the director of the Institute for Advanced Study. I was among many Princeton University professors who attended. Harish-Chandra, who was a professor of the Institute, caught my eye. He approached me and said, "Apéry proved such a great result, while you were lazily taking a nap. Now that you see somebody doing so wonderfully, you must think hard about your own work." This was a total surprise to me, as I had been friendly with him since around 1963. I responded, calmly and firmly, "That's a nice result, but I don't think it is so great. After all, many zeta values are even transcendental. A single irrationality result doesn't provide any new perspective." He made a very unhappy face, and mumbled something like, "But it was an old open problem ... ," the last part of which was inaudible.

Clearly he thought he finally found something with which he could humiliate me. To his disappointment, he failed. Did he do such a thing to other people? Unlikely, though I really don't know. But why me? To answer that question, let me first note an incident that happened in the fall of 1964. As I already explained, Atiyah and Bott proved a certain trace formula based on my idea. Bott gave a talk on that topic at the Institute for Advanced Study. In this case he clearly acknowledged their debt to me. In the talk he mentioned that Weyl's character formula could be obtained as an easy application. Among the audience was Harish-Chandra, who said, "Oh, I thought the matter was the other way around; your formula would follow from Weyl's formula." Bott, much disturbed, answered, "I don't see how that can be done." After more than ten seconds of silence, Harish-Chandra said, "It was a joke." There was half-hearted laughter, and I thought that his utterance was awkward and did not make much sense even as a joke.

It is futile to psychoanalyze him, but such an experience may allow me to express some of my thoughts. He was insecure and hungry for recognition. That much is the opinion shared by many of those who knew him. He did not know much outside his own field, but he was not aware of his ignorance. In addition, I would think he was highly competitive, though he rarely showed his competitiveness. From his viewpoint I was perhaps one of his competitors who must be humiliated, in spite of the fact that I was not working in his field. Here I may have written more than necessary, but my concluding point is: He did so, even though I did nothing to him.

Returning to the difficulties in understanding the significance of mathematical works, the problem is often caused by ignorance, or rather, the lack of mathematical sophistication. The above case of Harish-Chandra is certainly an example. Here is another good example unrelated to my work. *Basic Number Theory* by Weil published in 1967 is a textbook of algebraic number theory, whose principal features are class field theory and *L*-functions. There is nothing extraordinary, nor any new result. Around 1970 a well-known English mathematician, who worked on transcendental numbers, whom I call L here, was at the Institute for Advanced Study. He met Weil at a party given by Selberg, and vehemently protested against the title of the book. In his view, class field theory should not be called basic. Weil responded nonchalantly; I was merely a bystander and said nothing. Evidently L knew nothing about class field theory, and he felt most likely that the title suggested his lack of that basic knowledge. He must have known of the existence of the famous 1801 book by Gauss. Weil's book, and in reality class field theory, is a natural development of a major part of Gauss' book, but L was unable to understand that fact. There are many people on his level, and so there is no wonder that few people understand my work.

In my *Annals* paper of 1967 I obtained what I had been hoping for, but it took a few more years to extend the results to more general cases, and also to reorganize them in a better formulation. Though I knew that there were many more interesting problems in the same area, I became tired of continuing the work. Besides, some younger people took interest in my theory and began to investigate. Therefore I decided to do something new. In 1993 I wrote a letter to someone, who, I thought at that time, would understand what I wrote, but who turned out to be a mere opportunist. In it I explained the development of my ideas and also how I felt in 1972, which I can call one of the demarcation points in my mathematical career; in fact I later had a few more turning points. Here are excerpts from the letter:

"The next question was of course the zeta functions of those canonical models. There was also a question of finding $\ell$-adic representations for the forms of higher weights. I made a manuscript *An $\ell$-adic representation in the theory of automorphic forms* on the latter in 1968, but kept it unpublished. [Eventually it was printed in my *Collected Papers* vol. II, article 68c.] It was completed in Ohta's paper, which you know, I presume. As for the zeta function, I started to work on the case of $GL_2(F)$ with a real quadratic $F$. The problem is essentially to classify all two-dimensional abelian varieties over finite fields with $F$ as a subalgebra of their endomorphism algebras. In 1967–69 I found that the zeta function was of the type Asai later investigated (*Annalen* 226 (1977)). As he acknowledged, he did so on my suggestion.

"I never published my results on the connection of this zeta function and the arithmetic of the quotient algebraic surface. I merely presented them in a "secret seminar" in Princeton. I think Casselman was among the audience, but I'm not sure. At least he mentions this somewhere (his Corvalis article?).

"In the 1960s I was practically the only person who was working on these problems, except for a few younger people who participated in the project under my influence. Please note that Langlands started his program much later, after seeing my results.

"As mentioned above, the question for $GL_2(F)$ is to count the number of abelian varieties of a certain type. That is essentially so in a more general case too, and I could predict that I would not enjoy that type of mathematics. (The question for difficult, or nonstandard, arithmetic quotients requires more than that.) Besides, I was interested in various things, modular forms of half-integral weight, for example. The period of an abelian integral was another topic I kept in my mind since 1955. Therefore, instead of spending my remaining life by counting the number of abelian varieties, I decided to do something new. Also, Langlands became interested in the affair and I thought that my mission was completed.

"It is possible that you knew all these and still attributed everything to Langlands, but I think that is unlikely. Probably you knew vaguely that I developed the theory of canonical models, which people now call Shimura varieties, but I did so not only for their own sake, but also with the hope that eventually their zeta functions would be connected with automorphic forms, as in the cases I treated in my 1967 Annals paper. Of course Langlands knows this, as he mentioned it somewhere in his papers, though I don't think he ever mentioned any of my papers. [Langlands' article is "Some Contemporary Problems with Origins in the Jugendtraum," *Proc. of Symposia in Pure Math.* vol. 28 (1976), 401–404.]

"It is true that I never made precise conjectures in the most general case, but I don't think that is most important. For me, it is more important to show that there is a vast area in which one can really prove interesting theorems, and to give definite evidence by actually treating nontrivial cases. In fact, I am proud of having done so."

Let me add a few more comments. First of all, in the letter I mentioned the periods of abelian integrals. I had been interested in that problem since 1955, as I said in my article, "André Weil as

I Knew Him," which is included in this book as Section A5. I was able to obtain satisfactory results on this and published them in my papers in 1977–1980. In the letter I also mentioned that Langlands had become interested in my work published before 1972. I believe that he is responsible for the terminology "Shimura variety," and also that he never explicitly cited any of my papers. The reader who is interested in his reason for it may be encouraged to ask him.

Before closing this chapter, let me note here two examples of how I subconsciously found something, which may be compared with the suddenness of Poincaré's discovery of his new Fuchsian group. The first case concerns a family of special functions called confluent hypergeometric functions, which are important in applied mathematics. They appear even in number theory, and around 1979 while in Princeton I realized that I had to investigate their higher-dimensional generalizations. I was able to prove easy facts, but obtained no definite results, and so I turned to other subjects. One morning about two years later, I was making notes for my graduate course that afternoon. Suddenly what I was writing at that moment reminded me of that work left unfinished, and made me think, for no reason at all, that I would be able to handle the problem if I started that day.

After my course that afternoon, I took up the project again. Encouraged by the feeling that I would be able to do it, I could continue my work without interruption, and after a week I was certain that I had definite ideas, though it took six months to complete the work.

Here is another example of a different nature. Some time in July, 2001, I was giving a series of lectures at Kyoto University. Two years earlier in 1999 I had published a paper concerning the representation of an integer by a quadratic form, a topic completely different from the subject of my lectures. There was an old problem of finding the number of representations of an integer as sums of five or seven squares, which was mentioned in Hardy's book on Ramanujan published in 1940. Since then there had been no meaningful work on the question until 2001, nor did my 1999 paper settle it.

One day in a hotel room in Kyoto, after looking at the notes I had prepared in Princeton, I took a rest. Then I had a sense, for no reason at all, that I would be able to settle the problem about the sums of squares by the method of that paper, and so I jotted down a few ideas on my notebook. After a week, I left Kyoto for a much cooler place in Nagano Prefecture, and started to work on the question. Everything went smoothly, and to my surprise, I was able to solve that old problem within five weeks. The difference of this case from the previous one is that until that day in the hotel room I had had no intention of investigating that topic. I had only known that it was open.

Anyway, since 1972 I began to work in new areas; or may I say, I have succeeded in creating new areas of investigation. Also I discovered several old subjects which had been considered finished or forgotten, but actually had the potential of further development. I have been fortunate enough to be able to contribute some new ideas in those old areas of investigation. However, it is obviously unnecessary to talk about them here, and I end the description of my mathematical work at this point.

# CHAPTER IV

# A LONG EPILOGUE

## 18. Why I Wrote That Article

There is an article of mine "Yutaka Taniyama and his time, very personal recollections," published in the *Bulletin of the London Mathematical Society,* 21 (1989), 186–196. As I already wrote, I coauthored a book with him, and at one time I was constantly talking with him, but that period was not very long. I had known him since 1950, but I began talking mathematics with him only in 1954 and I went to Paris in 1957. He killed himself in November, 1958, and his fiancée did the same two weeks later. I was in Princeton at that time. The details of these are given in that article of mine in 1989.

Why did I write it? He was an unusually talented mathematician, but I did not write the article for the purpose of saying so. There is a section titled Taniyama's problems at the end, but I added it merely to comply with the editor's request. Ignoring that part, the last paragraph of the text may be taken as my official answer to the question of why I wrote it. But there was a more direct reason for why I wrote it *then.* I have never told it to anybody, and I very much hesitate to make it public, but here I summon my courage to write what I remember.

I knew his fiancée as I wrote in that article. Chikako knew him, and in fact we three had dinner together some time in 1957, which was about two years before our marriage. He once told her

G. Shimura, *The Map of My Life,* doi: 10.1007/978-0-387-79715-1_4,

that he enjoyed the movie *The King and I,* a fact I mentioned in the article.

One evening in early December in 1986, Chikako and I were having dinner and sitting face to face with a table between. We were talking about Taniyama, for what reason I don't remember, but that ended when our dinner finished. We were silent for a few minutes, and I was still thinking about him. Suddenly tears ran down my cheeks, which she noticed. She said, "What's the matter with you?" but I said nothing. Walking around the table, she came to my side, and put her hand on my back. Still I was silent. She then returned to her seat, and tears ran out of her eyes. We just kept weeping without a word.

Why did we weep? He was very much to be pitied, we thought. I say so, as I cannot find a better expression. The next day, driven by some invisible force I began writing that article, and finished the first draft within ten days. This is why and how it was written. In other words, I wrote it in order to keep back my tears. I am not sure the reader understands this, but I cannot write more than the above. Also this explanation may have been unnecessary for those who didn't read that article. Anyway I wrote here what I remember.

## 19. Impressions of Various Countries

Let me first add a few more things about France. I lost count of how many times I have been to France, but I remember that I have visited most of the notable places in Paris and its suburbs; I also spent a few enjoyable days in the Loire area touring some of the famous chateaux. I was never impressed by Versailles, which for me was merely an example of the monuments that represented the taste of the time. The Louvre in the 1950s was quite different from what we see now. Every room was packed fully. There were many glass cases each of which displayed hundreds of cylinder seals with no explanation or their impressions. There were many medieval paintings depicting the martyrdom of saints. Often four or more paintings of the same bloody scene were hung; it was as if the curator had no idea of selection. Possibly, they had no extra storage space, and so they displayed everything. That has changed of course, but I have a certain nostalgia for the Louvre in the old days.

There is a building called Le Louvre des Antiquaires, just north of the museum, which houses many antique shops. Since its establishment, which I think was in the 1970s, I found myself going there more often than to the museum, as I was able to have a closer view of the displayed objects, and even to touch them. Moreover, I could have been able to own them if I had really wanted to, but I did buy only a small Imari bowl of a passable quality.

Because of my fondness for decorative arts, I visited the Museum of Decorative Arts in Paris and also the National Porcelain Museum in Sèvres. In an antique shop a few days earlier, an eighteenth-century cup and saucer set caught my eye, but I didn't buy them. In the porcelain museum I noticed a set which was exactly the same except that the cup had a visible chip. This made me think of returning to the shop, but at the end I persuaded myself not to become too obsessed by the idea of owning such an object.

In 1987 Chikako and I lived in an apartment close to Place d'Italie for two months. Since the coffee cups we found there were not of our taste, we decided to acquire something that suited us, and went to rue de Paradis, a street lined with many shops selling dinnerware. However, the designs we found there were uninteresting; just small flowers or sprigs spread all over the surface of the piece, like women's pajamas. I then realized that the French housewives were so conservative that they could not accept anything else. That kind of conservatism had prevented the Impressionists from succeeding, but the same kind of oppression seemed to exist in the decorative arts. The progressive boldness in paintings we have seen in French art didn't exist in dinnerware. At the end we bought two sets of cups and saucers tolerable to our taste, as a souvenir of our defeat in a battle with French conservatism.

There were many fish shops near that apartment, and we were able to find raw sea urchin at one of them. One morning Chikako said to me, "I'm going to Galeries Lafayette," and departed. I was working at my desk, and simply said, "Hmm." After a while she came back and showed me a large portion of fresh bonito including its head, which she bought on the top floor of Lafayette. This time she was not short of money, and was able to settle her old scores of Osaka in Paris. That incident impressed me because in Japanese department stores such things are sold on the first or second basement floor. She still had no silk robe; she merely sewed a skirt on the machine that was in the apartment with a piece of cotton bought at Bouchara, a fabric shop near Lafayette.

We encountered three pickpockets in that period. Once at a subway station, Chikako was searching for a bill in her wallet to buy a ticket, when a man snatched the wallet and fled. He set his feet on the first step of the escalator and held the wallet above his head as if he was showing off a trophy. My wife, known in our family for being fleet of foot, ran after him. As the people in front of him slowed him, she was able to reach him and to snatch back the wallet, to everyone's surprise. A similar incident happened to me, when I was holding a bill of a large denomination for the same reason. A pickpocket snatched it, and because I did not have the quick reflexes of Chikako, he was able to escape. There was one more encounter, in which the pickpocket tried his art unsuccessfully, and so our total loss was that bill.

The International Congress of Mathematicians of 1966 was held in Moscow. I gave a talk on the contents of my paper dedicated to Weil there. I was glad to have a chance of seeing that communist country. I had an enjoyable time as a tourist, but otherwise I ended up with the conclusion that the Soviet Union was extremely strange and even detestable. First of all, the proceedings of the meeting were published as a booklet of less than 200 pages, which was unusually meager compared with those on the previous occasions in other countries. As for the text of my talk, they sent me a nine-page proof, and I sent it back with my corrections. I later discovered that they made many changes, eliminated many lines, and did not make corrections as I indicated. This was the only time in which my manuscript was shortened substantially after my correcting the proof.

The city looked very poor; there was little merchandise in the main department store. Our family of three stayed in the Hotel Metropoli, which at one time was one of the most prestigious hotels in Moscow, but had not been renovated for a long time. The bill for each of our meals was almost always incorrect. They did not seem to be cheating; probably they were either casual about such matters, or incapable of being precise.

In 1991 there was a conference in Minsk, Byelorussia. My wife accompanied me, and we were assigned a suite in an old hotel, which had a kitchen and two bathrooms, and could be called an apartment. Though it was in late May, it was still cold in that city, particularly at night, and the room was not heated enough; besides, no hot water was available. We reached Minsk by train from Moscow, where somehow I lost my collapsible umbrella. Since it was raining, I told my problem to one of the local organizers, who brought me to the dollar shop in the city, where I bought a replacement for ten dollars. The place was open only to foreigners. Where and how could ordinary citizens buy such everyday goods? I had no idea, except that I noticed about eighty people queueing up for the sale of bedsheets which was being held on a streetcorner. Such sales seemed standard even for fresh vegetables, as I saw a queue of about forty people intending to buy cabbages and tomatoes.

The wife of one of the participants of the conference had to send packages from Minsk to Moscow. I had nothing to send but was curious about how the matter would be handled, and so I accompanied her to the post office. One would buy a wooden box there, put the material in it, nail down the lid, tie up the box with a string, cover a knot in the string with waxlike stuff, and press a seal on it. Since there were only two hammers in the room, the senders had to wait for their turn.

When we dined at the Hotel Metropoli in 1966, we often had caviar and champagne, simply because they were easily and naturally available on the menu, not because we wanted to eat in a luxurious fashion. Indeed almost every guest was doing the same. There were no such things in Minsk. Guided by the French description of the items on the menu, we would order something, but what we got was always different from what we expected. Maybe we were supposed to be entertained by a surprise at each meal.

After the meeting in Minsk, we went to St. Petersberg, which was still called Leningrad then. I gave a talk at the institute there. Four or five people listened to my talk, and in addition two people were conversing loudly in the back of the lecture hall, which did not bother my audience, and so I pretended not to be concerned. We were assigned to a shabby old-fashioned hotel, whereas most foreign tourists seemed to stay in modern well-appointed hotels. At that time one dollar was equivalent to eighteen roubles, and we often met on the main street someone who would be happy to buy a dollar for forty roubles, but I always declined the offer.

We went to the tourist office to purchase tickets for our sightseeing trips. They demanded a high price in dollars, for the reason that we were foreigners. I countered by saying that I was a participant of a conference sponsored by the Soviet Academy, and received expense allowances in roubles, and showed the documents as the proof. Then they reluctantly sold tickets to me with a lower price in roubles. Incidentally, the Soviet government collapsed within that year and Yeltzin became the top person in charge.

I visited China, another communist country, in 1984. I had been acquainted and friendly with several Chinese mathematicians who came to Princeton, and they were able to arrange my visit. Also, I had been familiar with classical Chinese literature, and had wished to go there at least once in my life, and so was very happy to fulfill that wish. Though I was treated graciously and I had no bad feelings like what I had against Soviet Russia, at the end of my visit I was left with the impression that the country was full of irreconcilable contradictions.

The hotel where I stayed in Beijing was built in an old Russian style, and there was a key-bearer on each floor, as was so in the Hotel Metropoli; the meals were Chinese. There were two types of bank-notes: one domestic, the other for foreigners. I was told that every household had a small television set, whose price was five times as much as the average monthly salary of a factory worker. Incredibly many bicycles were running on the streets, but they were of a single-speed type; multiple-speed ones had just appeared on the market a few months earlier.

The buildings for the mathematics department of the university were old-fashioned and reminded me of those of the University of Tokyo in the 1950s. Wagons pulled by donkeys were moving slowly on highways side by side with automobiles. Roadworks of the simplest type were being done by human hands with no machines, as if that was the government's idea of increasing employment.

After ten days in Beijing, I went to Xian, where I saw the famous clay soldiers of Qin Shi Huang. The city looked far poorer than Beijing; indeed, often a single donkey was pulling a small wagon, whereas in Beijing three donkeys were used for a larger wagon.  There were people on the roadside selling vegetables, whose quantity was so small that not much money would be earned even if everything was sold.  From Xian I went to Shanghai by train which took about twenty-four hours. There were two classes, called soft seat and hard seat. My translator-guide and I had soft seat tickets. There was a great difference between the two classes, and the discrimination made me uncomfortable, though of course I was not responsible for it.

A warning against spitting was posted everywhere in every city, but I saw only one instance of spitting by an old man during my stay in China. In Su-zhou, to call a taxi, my local guide spoke Chinese instead of English, which made the taxi company think that they would be unable to earn foreign money, and they let us have, instead of an ordinary taxi, an autobicycle to which a three seat carriage was attached, the type of transportation gadget that I had never ridden in before or afterwards. Because of this, however, I was able to make an interesting excursion among ordinary Chinese citizens.

In Xian and Shanghai, I talked with several students, whose lack of basic mathematical knowledge disappointed me. The university bookstore at Shanghai was full of exercise booklets of English examination problems for studying in the United States, but had nothing else in foreign languages.

I recalled that in my high school days I read many pirate editions of books in English and German, which were being called Shanghai editions, and I would think were most likely printed there. I guess no such things existed in 1984. It was also my impression that Chinese students did not have easy access to the latest mathematical publications in foreign languages.

I returned to Tokyo after three weeks in China with various uneasy feelings, especially about the universal poverty there. After all, I once lived in deprivation for more than ten years during and after the war, and so was able to empathize with those who were suffering. Twenty-three years have passed since then, and I have no precise information about the present situation. But I doubt that mathematicians are well-respected or well-treated in that country now.

Turkey is my favorite foreign country, where Chikako and I have been five times, though one of our visits was only an eighthour sojourn in Istanbul. We went to that country for the first time in 1987, when we were staying two months in Paris. I had long kept my wish of visiting Turkey, and having realized that it took only three hours to reach Istanbul, I decided to devote the spring break to our trip, with no definite plan except for a one-week hotel reservation in the city. We mostly stayed there, and took a full day bus tour to Bursa. We enjoyed our trip beyond our expectations, and as a result we returned to that country repeatedly.

Let me mention here a few things which we fondly remember but are not noted in the guidebooks. There is a famous bridge called Galata, that connects the old section of Istanbul to the new section. It has two tiers; the upper tier is for automobiles; the lower tier is for pedestrians and restaurants. One morning, as we were walking across the bridge, a restaurateur stopped us, showed us the fish he was grilling, and urged us to eat. Since it was too early for lunch, I said so and we kept walking. He then called from behind us, "Bakayaro, bakayaro," in Japanese with a funny accent, which could be translated, "You fool, you fool."

There is a fairly large mosque called Yeni Cami (New Mosque) just south of the Galata Bridge. We entered it through the door on the north side. After appreciating the architecture of the inside of the mosque, which was as good as any average mosque, if not extraordinary, we could exit through the door on the south side. There we found a small square which functioned as an open air market consisting of tea stalls and tiny shops selling flowers, birds, and small animals. The west and south sides of the square formed an arcade called the Egyptian Market, lined with many stores that sold spices, herbs, nuts, grains, and kitchen utensils. When we came to this square for the first time, after walking through the dimly lit interior of the mosque, we were struck by its charming view under the bright sunlight, which was totally unexpected, and left an unforgettable impression. So we came back to this place each time we visited Istanbul.

There is a district called Kum Kapu in the southern section of the city facing the Sea of Marmara. A few blocks north of the seashore is a small square, with a fountain in its center, surrounded by many restaurants known for fish dishes. We were staying in a hotel that was ten-minute walk from the square, and so we often went there for dinner. We would walk a back street in the evening and observe some men playing a game which looked like dominoes, and also would find Turkish-style tombstones in a small graveyard adjoining a mosque. The area was rather untidy and meager, but had a certain warm and homely atmosphere that reminded me of Ushigome in my childhood. Though our primary purpose was to eat at some restaurant, I definitely liked walking on streets with such a feeling.

In 1992 we spent April and May in Ankara. I gave a course at Middle East Technical University in the city. The high rate of inflation in Turkey is well known, but most short-term tourists do not notice it. In Ankara we found a confectionery shop selling marrons glacées, far less expensive than in the States. We bought a box there almost every ten days or so, and found that the price was raised each time. Of course one could say that it was the same in terms of the dollar. A menu would be posted at the door of every restaurant, but once it disappeared at one of the restaurants where we often ate. We thought that the place was closed that day, but it turned out that they had no time to post a new menu with raised prices.

We also went to Trabzon, a city on the Black Sea coast founded in the seventh century B. C. There is the ruin of a Byzantine monastery built in the hollow of a high cliff, about thirty-five miles west of the city. Our hired taxi followed a narrow road along a ravine that flowed between hills covered by lush damp greenery, which we had not seen in Ankara or Istanbul. We were often greeted by a large group of sheep that appeared on the road led by a herdsman. Each time our car stopped and waited until they crossed to the other side. In Trabzon we had various kinds of interesting experiences. I narrate here just one of them. We saw a couple of men on the street, who were glancing at us from time to time and discussing something. One of them approached us and asked in Turkish where we came from. At that time I spoke a little of that language, and answered that we were Japanese. He then went back to the other man, triumphantly shouting something. Clearly he won the bet about our origin.

Wherever we went, we heard the words from a minaret of a mosque, which announced the time for prayer. We very much liked that tune, which sounded more soothing than exotic to us. In fact I prefer mosques to churches. I always felt a certain warmth in a mosque, and was tempted to stretch myself out on the carpets spread on the floor, whereas the cold stone structure of a Catholic church put me off. Apparently Turkish intellectuals have mixed feelings about such religious sounds. I, as a foreigner, have the privilege of expressing any subjective and irresponsible opinion. Once I had a friendly conversation with a taxi driver, who said to me that he believed in God, and added, "I'm sure you do too, don't you?" I responded "Yes, I do, but my God may be different from yours," which made him silent. But that is irrelevant to my reaction to the sounds from the minarets.

However, even as a foreigner, I was concernedly impressed by the large number of shoeshine boys in Turkish cities. It would be desirable to prohibit labor by minors, or to lengthen compulsory education, I thought. I wonder if there have been any changes in that respect since then.

In Ankara we found a street lined with many small shops brightly lit by many electric bulbs that sold gold accessories, such as bracelets, necklaces, and brooches. Trying to find a souvenir, we went into some of them, and discovered that those shops were all alike and were selling almost exactly the same merchandise. We picked one item, whose price the dealer set after weighing it on a scale. We later found similar clusters of shops in Beirut and Tripoli, both in Lebanon, and also in Cyprus. It is believable that such shops are common in the Middle and Near East. All the cleaning ladies of our apartment in Ankara, without exception, wore gold bracelets. Probably they preferred gold objects to bank accounts as their way of saving, which seemed to explain the raison d'être of those shops.

From Ankara we traveled to the northern section of Cyprus. We visited the castle at Famagusta, which was supposed to be the setting for the opera *Otello*. One evening at a hotel in Kyrenia, a northern port city of the island, we witnessed a wedding ceremony. The bride and bridegroom were standing in the center of a hall adjoining the lobby, and there was a long line of people, who we thought were expressing words of congratulations and best wishes to the couple. We had never seen that type of event anywhere else.

We encountered another type of interesting wedding incident a few weeks earlier in a city south of Ankara. We saw a wedding procession of a dozen people on a road, when a bus came from the opposite direction, and stopped in front of the people. The passengers of the bus, who looked like high school girls, all got off, circled the bride and bridegroom, and began dancing and singing. After a few minutes they got on the bus, which went away, continuing its original course.

Before returning to Princeton, we went to the post office to mail a few parcels. The system was similar to that in Minsk, though plastic boxes were used instead of wooden ones. No nails and hammers were necessary. The boxes were sealed properly, and what we sent arrived home safely.

We visited Lebanon twice in 2000 and 2004. We stayed in a hotel close to the American University of Beirut, where I gave talks both times. The hotel was new in 2000, and so we were asked to take our breakfast at a nearby restaurant. But it was closed on one Sunday, and the hotel receptionist directed us to another restaurant two blocks away. We found it, sat at a table, and ordered coffee and croissants. The waiter said coffee was available, but croissants were in a locked room, and he did not have the key. Seeing our unhappy faces, he said, "I think I can get them for you at the supermarket over there," and went out. We were eventually able to have our breakfast as usual. In fact, we had heard, "I can get them for you," already at a restaurant in Trabzon. I used to buy *The International Herald Tribune,* whenever it was available in the country I visited. One day in Beirut I bought a copy from a street vendor; it turned out to be that of the previous day.

In Lebanon we enjoyed a few trips to famous places of scenic and historic interest. From Beirut we also went to Damascus in Syria and had interesting experiences, but it is clearly too much to narrate all of them, and so I end this section by writing about Iran, where we went in 1994. One of my friends who came to the States from Iran kindly arranged a plan for me to participate in the yearly meeting of the Iranian Mathematical Society held in Teheran. We were urged to take a sightseeing trip to Isfahan before the meeting. It was March 24 when we reached there, and the Iranian people were celebrating their New Year in a warm climate, comparable to that of early May in Princeton.

A river flowed through the city and there were open fields on both sides, covered by grasses. We saw many families picnicking there. A husband and a wife were grilling some pieces of meat on a portable stove. When it became the time for prayer, all those in the fields including the couple knelt with their heads touching the ground. But the wife raised her head from time to time, and turned over the pieces of meat. "Isfahan is half the world," is a phrase often heard, and we were able to admire the beauty of many of the famous monuments of the city, and perhaps we saw one-fourth of the world.

As for the meeting in Teheran, each morning session was preceded by the projection on the screen of the Iranian national flag fluttering in the wind, with a music piece played through the microphone, which I presumed to be the national anthem. Some taxi drivers in Teheran told us that they had worked in Japan for a few years, and had enjoyed their stay. Many citizens we met fondly mentioned *O'Shin,* a television show imported from Japan, whose eponymous heroine had captured their imagination. The taxis we rode in Teheran had no meters; the driver casually said 2000 ryals

The Shimura family at the Moscow amusement park. From left
Chikako, Tomoko, Goro. August 1966

See the description on page 160 (first two lines of the second paragraph). Teheran, March 1994

or 3000 ryals, or whatever price he thought appropriate, depending on the destination.

After four days in Teheran we returned to Princeton. While in Iran, Chikako wore a raincoat and covered her head with a scarf all the time except in our hotel room; it was so even when she was eating. We were the only guests in the hotel restaurant on the last evening before our departure, and three waiters came to our table and chatted with us. Asked whether she liked the country, she responded, "Yes, very much, except that I don't like to wear the coat and scarf constantly." One of them then said, "It's a matter of several more hours of patience, until you board the plane." Another said to me, "I'd like to work in Japan, What do you think of the prospect?" "I don't really know, but the Japanese government seems to be setting a higher hurdle for it," said I.

The photo on page 159 shows Chikako at Hazrat-e Abdol-Azim Mausoleum, the most important pilgrimage spot of Teheran. Here the scarf and raincoat were not enough; she had to wear a hooded gown provided at the gate of the compounds.

Three months later, we met a young Iranian woman in Princeton. She also told us that *O'Shin* was popular in her country, and Japan was being called the country of *O'Shin*. Mohammed had a daughter who died young. There was an Iranian television program on her birthday or death-day, and the host asked a group of Iranian girls who their model female personage was. Some mentioned Mohammed's daughter, and some O'Shin. Later, it is said, all those involved in producing the program and those who gave the wrong answer were punished. Also, Ayatollah Khomeini was supposed to have condemned O'Shin as a girl not obedient to her parents.

## 20. You Will Find Out

One day at a party in the fall of 1962, I told André Weil the following opinion of mine. "If someone wishes to be a first-rate mathematician, then he gradually elevates himself and reaches a certain height at the age of perhaps 40 or 45; after that he should try (or he must be able) to keep that level as long as possible. That is my ideal." Then I asked him what he thought. He merely said, "You will find out." At that time I took it to be negative.

Around 1980 I reminded him of this conversation, which he remembered. He said that he didn't say it negatively, and added that "Hardy's opinion that mathematics is a young man's game is nonsense." I am inclined to think, however, that he was not completely sure of himself in 1962, when he was 56. In other words, hearing what I said, he must have had mixed feelings, and evaded the issue. After all, he could have said, "That is a good idea," or something of that sort, but didn't. Maybe he found out himself.

I recall one more conversation with Weil. Four volumes of the works of Hermann Weyl were published in 1968 (*Gesammelte Abhandlungen* I–IV). I noticed them in the library of the old Fine Hall, and was surprised by the abundance of minor papers in his late years. I told this to Weil, who said, "Perhaps he tried to be useful to the mathematical public in his own way." This may have been a reasonable answer, but sounded like a quibble to me. As the reader might have already noticed, I tend to be harsh regarding the self-importance of older mathematicians. I don't understand why they are willing to produce insignificant papers or to make haughty remarks with no substantial mathematical content. I should mention, however, an exception: Siegel did much better.

But there is a more troubling aspect in this matter, not in what Weil said, but in what Hardy actually said. In Section 4 of *A Mathematician's Apology* he writes:

"No mathematician should ever allow himself to forget that mathematics, more than any other art or science, is a young man's game."

I found this rather pompous. Why did he insert the words "ever allow himself to"? Perhaps he thought those words would make his statement more dignified. Also it is arrogant and wrong. If he had said, "For me, mathematics is a young man's game," that would have been acceptable, but he demanded every mathematician not to forget that mathematics is a young man's game. Suppose someone forgets it or even has never had such an idea. What's wrong with that?

Hardy wrote it when he was 63. He felt his career as a creative mathematician was over, which was true. Therefore he invented such a general principle of a young man's game in order to defend or justify his demise as a mathematician. I haven't heard any opinion about it except what Weil said, but I would think Littlewood, his principal collaborator, never took it seriously. He was not as affected as Hardy.

Reading his *Apology* and the "Foreword" to it by C. P. Snow, we have a fair picture of him as an extremely competitive man. Apparently he was always worried about his ranking among contemporary mathematicians. That is unusual. Probably Littlewood was not such a type, as most mathematicians are not.

I can think of only one other mathematician comparable to Hardy in that respect. At a dinner party in the mid-1960s, I sat next to Don Spencer, talking about the odd behavior of a young mathematician. I told him casually without much thought, "Maybe he needs recognition and that will help," to which Spencer immediately replied, "No, that won't work. Once he is recognized, he will need more recognition." This puzzled me, as I was unable to agree with that opinion. Later I realized that he had a certain individual in mind: Norbert Wiener. Spencer knew him at M. I. T. and apparently their relationship was strained. In any case, Wiener was known to be a person who kept asking his colleagues, "Am I slipping?" I heard a more elaborate version of this utterance from a few people.

There is a more interesting case of a philosophy professor who taught at a well-known private university in Kyoto from 1930 to 1960. A visiting German philosopher once asked him, "Who do you think is the best philosopher in Japan now?" to which he replied, "That's me." Somewhat taken aback, the German scholar said, "In that case, who is No. 2?" After thinking a while, the professor answered, "There is no No. 2."

I add here what a well-known Japanese mathematician said. After teaching at three American universities, he returned to Japan, and taught ten years at the University of Tokyo. A few years before his retirement at the age of 60, he said to his younger colleagues something which could be condensed to: "We professors are here to teach, not to do research." This was told to me by one of those young men, who was much disturbed by these words. It is likely that the professor wanted to justify his inactivity at that time.

Returning to "my ideal," I indeed thought so at that time, but what "I found out" is somewhat different, and I now feel that I should have said that "A mathematician should continue making progress" and I was fortunate enough to have been able to continue making progress, at least from my subjective viewpoint.

Almost all famous composers kept making progress in their careers. Mozart composed great music in his early twenties, but we can clearly see his advancement to higher levels in his late years. There are many similar examples of painters or novelists whose later works are much better than earlier works. Why not mathematicians? That is what I am saying, but there is one thing which distinguishes mathematicians from other professions in art. Composers, painters, and novelists have the general public as their audience, and are therefore always judged by the latter's evaluation. If an artist's work is not appreciated by the general public, he cannot survive. On the other hand, there were mathematicians and scientists who spent their late years merely basking in the fame they achieved in their youth.

Odilon Redon began his career by creating weird drawings of phantomlike objects, but had no large audience. When he lost the source of income on which he had been relying, he changed his style completely by producing more colorful pictures of flowers. It may be a matter of taste, but I would call it his progress, as he went in a new direction.

Indeed, many mathematicians, after obtaining a doctoral degree, merely continue their work in the same narrow area. That is their choice, and I am not in a position to advise them. I would never say "No mathematician should ever confine himself in the domain he had chosen when he was young." Still I always wonder why they don't try to go in different directions. The matter depends on one's personality. In any case, I think that the effect of one's temperament is stronger in one's development than that of one's talent. Whether one can find a new direction by throwing away a fixed idea is principally due to one's temperament, not to one's talent.

## 21. The Other Side

Once I had an extremely enjoyable day, which I always considered one of the happiest days of my life. It happened when I was a student at the University of Tokyo or possibly just after graduation. At that time there was a well-known Kyogen (Nō comedy) actor named Nomura Manzo, who had two sons, Taro and Jiro (later Man-nojo and Mansaku), both of whom were also Kyogen actors. Normally a Kyogen is played along with a serious Nō play, but the Nomura school would hold, once a year in the spring, a program that consisted only of Kyogen plays. Somehow my brother was able to obtain tickets. I think he and I attended at least three consecutive times. Normally five Kyogen plays were presented, but once ten were staged, five by professionals and the other five by amateurs assisted by professionals. I view Kyogen as one of the most refined drama forms in Japan, and so I had been enjoying every performance, but on that day I was able to see ten performances, all of which were superb. But such a day never occurred again, though I enjoyed Kyogen later on various occasions, even in Princeton.

Let me now describe different kinds of happy experiences I had and their aftereffects. As I already mentioned, I went to Boulder, Colorado in the summer of 1963. I drove from Princeton to Boulder in late June with my wife and daughter who was three years old. In Colorado we enjoyed seeing many interesting things that could not be found on the east coast. We would naturally go to the Rocky Mountain National Park; we would also drive to the points near the tops of several high peaks more than 14,000 feet above sea level.

After spending an extremely enjoyable summer, I started the fall term teaching at Princeton. I would go to my office in the old Fine Hall, and would be doing something at my desk. Then I would recall the happy summer days of just a few weeks ago, and would ask myself why I was there doing stupid things instead of being in Colorado. That was a strange feeling of the kind I had never experienced until then. Of course, occasionally I had recalled happy experiences in my past, but it was for the first time in my life that I compared the present with the past in such a negative manner.

For me the United States in the early 1960s was a place full of hope, and I feel as if those days were glittering in gold. It took five days to reach Boulder. While we were eating at a motel restaurant, other drivers would approach us and would ask various questions. One of the most frequently asked questions was whether we intended to become U.S. citizens. They were always kind and friendly. It seems that the people of the United States have become more callous and less relaxed these days compared with forty years ago.

I have a special reason for writing such experiences. First of all, being a mathematician, I belong to the type of people who always *think*. But there is another type of people who *do not think*. Of course that is an exaggeration. But there are those who are carefree and happy, and satisfied with the present without thinking. Let us call them the inhabitants of *the other side*, and call my side *this side*. It is difficult to define both sides more accurately, but such is unnecessary. It is sufficient for the reader to have some vague notions of what I am writing.

The people of the other side have their own troubles, but they are of a different kind from mine, and that is the principal reason for their happiness. Besides, since I am watching them, they can be called those who are watched, whereas I am a watcher. Though I am envious of them to a certain degree, the envy is not the most important factor. Still, I wished I had been on the other side.

In any case I had that idea of this side and the other side for some time since my late teens, though I cannot say that I was always conscious of it. Then at the age of thirty-three I became someone who other people thought was very happy. I had a similar experience later when we went to Canada. Apparently people viewed my family as immersed in an enviable happiness. Of course I cannot tell what they really thought, but I felt that oh, at last I reached the other side, which was true in the sense that I became a "being watched." I did not think I had attained full citizenship of the other side; one might say that my application for it was being processed.

Looking back on one of my preschool days when the young and pretty woman opened the door and invited us children into her garden, I think we were on the other side from her perspective. Or rather, she and we were all on the other side from my perspective at a later date.

There is a scenic park in Paris called Parc des Buttes Chaumont, where ordinary tourists rarely go. Unlike other parks in Paris, this one has many ups and downs; there are steep stone steps from the top of a hill leading to a pond. There used to be a ferryboat at the shore of the pond, as if it had been waiting for a fleeing couple who were trying to evade their pursuers. There are also flower gardens and large lawns as in other parks in Paris, but overall this park has a unique atmosphere which cannot be found in any other European park. When I was in Paris for the first time, I went there many times and cherished its unusual features. Naturally I boarded the ferryboat, paying a fare, which was fifteen francs. But being alone in a park made me feel that something was missing. Therefore I hoped someday I would return to this park with my future wife. It must be this park, not any other park.

While in Paris, there were three proposals of marriage to me. One of the possible mates was French; I met her and also a Japanese girl; I knew the remaining one who was Japanese only by her photo. Nothing came about, as somehow I was unable to be persuaded. In other words, I had no urge to play the role of a man running away with any of them on the ferryboat.

Eight years later in 1966, I, with my wife and daughter who was six years old, spent five days in Paris. We of course visited the park one fine day in late summer. Nothing had changed since my last visit. The ferryboat was still there, and we boarded it. Thus my wish came true. I was happy to make my companions happy, but otherwise I had only a lukewarm satisfaction; I had been expecting much stronger feelings.

Reflecting on this matter later, I understood the reason for my reaction. I already had those unforgettable experiences in Colorado, and what happened later became inevitably a mere revival of an old play. Namely, I should not have expected too much. Perhaps in my subconsciousness the ferryboat had been the symbol of the vehicle that would transport me to the other side. If so, my experiences in Colorado had made the boat lose its symbolic meaning. However, my wife, without knowing what I was thinking, always talked fondly about the Coquilles Saint-Jacques she had at lunch that day in a restaurant near the entrance of the park.

# APPENDIX

## A1. That Conjecture

The title refers to my conjecture that every elliptic curve defined over the rational number field can be uniformized by modular functions. I told this to J.-P. Serre and A. Weil in September 1964; the incident is well recorded and known to many researchers. The statement is now a theorem, as it was proved some thirty-odd years later. On the other hand, Yutaka Taniyama made a statement in the form of a problem in 1955, which has a certain connection with my conjecture. It is my impression that practically no mathematician, except me, understood exactly what he said and its difference from what I said. Therefore in this section I explain these points in detail. I also think that there are many mathematicians as well as nonmathematicians who would like to hear my clarification once and for all.

First, an elliptic curve is a curve of the form

$$E : y^2 = x^3 + ax^2 + bx + c,$$

where the right-hand side has no multiple roots, that is,

$$x^3 + ax^2 + bx + c = (x - \alpha)(x - \beta)(x - \gamma)$$

with three different numbers $\alpha, \beta$, and $\gamma$. Here $a$, $b$, $c, \alpha, \beta$, and $\gamma$ are complex numbers. We say that $E$ *is defined over the rational number field* if $a$, $b$, and $c$ are rational numbers. To make our terminology short, let us hereafter call such an elliptic curve *rational,* though this is not standard. We do not assume that $\alpha, \beta$, and $\gamma$ are rational numbers.

Why do we consider such a curve? There are many reasons. Simply said, it is a natural object of study that appears often. Besides, if one wishes to investigate more general types of curves, one first has to know the properties of the curve in such a special case, which are actually very interesting and not so simple.

Next, what is the meaning of the expression "uniformized by"? Take, for example, the curve $C : x^2 + y^2 = 1$, which is a circle, not an elliptic curve. For every real number $t$, the point $(\cos t, \sin t)$ is on $C$, and conversely every point on $C$ can be given as $(\cos t, \sin t)$ with a suitable real number $t$. We express this fact by saying that the circle $C$ is uniformized by the trigonometric functions cosine and sine.

Suppose we can find two functions $f(u)$ and $g(u)$ of a variable $u$ on some domain $U$ satisfying

$$g(u)^2 = f(u)^3 + af(u)^2 + bf(u) + c.$$

Then the point $\big(f(u), g(u)\big)$ is on the curve $E$. If every point on $E$ can be obtained in this fashion, we say that $E$ *is uniformized by the functions $f$ and $g$.* In this case, for some natural reasons, we have to take $u$, $f(u)$, and $g(u)$ to be complex numbers.

There is an important class of functions called elliptic functions, by which every elliptic curve $E$ (not necessarily rational) can be uniformized. However, our problem is not uniformization. We are interested in arithmetic properties of $E$. Without going into details, let us simply say that given a rational $E$, we can define some function $Z$, called the zeta function of $E$, that embodies the essential arithmetic properties of $E$, and it is important to determine $Z$ for a given $E$. Now the uniformization of $E$ by elliptic functions can be employed for the determination of $Z$ in some special cases, but useless in most cases.

In 1954 Martin Eichler proved a result on algebraic curves uniformized by certain modular functions. This class of curves included rational elliptic curves whose zeta functions could be determined. This result was an important step, but not completely satisfactory, since there were only finitely many such elliptic curves, whereas infinitely many rational elliptic curves exist. Towards the end of 1954 I took up this subject, and started my work on this by constructing the arithmetic theory of modular functions in my own way. By July 1956 I was able to obtain a more general result than Eichler's, which was the work published in *Comptes rendus* of 1957 mentioned in Section 13.

Investigating further in the same direction, I found more rational elliptic curves for which $Z$ could be determined; also I found a good perspective of arithmetic uniformization of algebraic curves by modular functions. Though I had this theory already in 1964, or possibly earlier, I gave a full exposition only in 1971 as a chapter of my book published then. Now my conjecture means that every rational elliptic curve can be obtained in this way.

Before proceeding further, I must note a few facts known to every serious researcher in 1955. The zeta functions of some rational elliptic curves had been determined by that time, due to Deuring and Eichler. Looking at those known cases, one could easily guess what kind of analytic properties the zeta function of a rational elliptic curve must have. In 1936 Erich Hecke showed that a function with such analytic properties must be related to automorphic forms of a specific type. Modular forms are included in this class of functions, though Hecke's theory cannot distinguish them from other "useless" functions.

In 1955 Yutaka Taniyama made a statement as a problem, which was reproduced as Problem 12 on page 9 in my *Collected papers,* vol. IV. There, he considered an elliptic curve over an algebraic number field, which is clearly wrong; it must be restricted to a rational elliptic curve as I am considering now. After making this correction, what he said can be rephrased and condensed as follows:

*Given a rational elliptic curve and its zeta function $Z$, if $Z$ has good analytic properties (as Hasse conjectured), Hecke's work connects $Z$ with an automorphic form $f$ of weight 2 of a special type. If so, it is very plausible that $f$ is an elliptic differential of the field of associated automorphic functions. Therefore, given a rational elliptic curve $E$, is it possible to show that its zeta function has good analytic properties by finding a suitable automorphic form?*

Some mathematicians took this as a kind of conjecture, but there are a few nontrivial points that make the statement problematic, and in fact meaningless. That Hecke's work connects $Z$ with an automorphic form $f$ of weight 2 of a special type is true as I explained above, but the next statement does not make sense.

Hecke's functions depend on a parameter which he denoted by $\lambda$. If $\lambda \leq 2$, no nontrivial automorphic form (or more precisely, cusp form) corresponding to a differential form will appear, and so we have to assume that $\lambda > 2$. In this case the space of automorphic forms is infinite-dimensional, and the field of automorphic functions is not a field of algebraic functions of one variable; also the associated Riemann surface is not compact. That an automorphic form of weight 2 defines a holomorphic differential form on that noncompact Riemann surface is trivially true. But how can one call it an elliptic differential? In other words, how can one associate an elliptic curve with the automorphic form? There is no way to do so, as the Riemann surface is noncompact. Thus Taniyama mentioned Hecke's paper, but he said nothing meaningful beyond that.

If we restrict automorphic forms to modular forms, then we can do something, as I indeed did in my work. But that was not what Taniyama was thinking. In fact, he said, "Modular functions alone will not be enough. I think other special types of automorphic functions are necessary." (See ibid. page 10.) He was wrong here.

It should also be noted that though the theory of modular or automorphic forms is an old subject, its importance in number theory, particularly in connection with algebraic curves became known to a wide mathematical public only since 1964, when I gave a series of lectures at the Woods Hole conference, as mentioned in Section 17.

I can mention the following names as those who, in 1955, were familiar with the notion of the zeta function of an elliptic curve and also with Hecke's paper of 1936: Taniyama and myself in Japan; Deuring, Eichler, and Weil outside of Japan; otherwise nobody in the United States. Artin was capable of understanding what was stated, but I would think his knowledge of modular forms was minimal. As for Weil, he said something to the effect that automorphic forms would help, but only to a limited extent, and so he was definitely behind the times.

Speaking about my work, I had been constructing the arithmetic theory of modular functions in my own way, and always thought that Hecke's functions other than modular functions were useless for the problem on elliptic curves. Therefore I did not think much of Taniyama's statement, nor did I discuss this question with him. I co-authored a book with him on complex multiplication of abelian varieites, but it is on a different subject and is irrelevant.

There is one more point. From 1955 through 1964, I was the only mathematician who was investigating the zeta function of an algebraic curve uniformized by automorphic functions. Therefore I doubt that Taniyama's statement inspired any meaningful mathematical work.

So much for the historical facts. It is my opinion that anybody who wants to say something about this conjecture should first understand the precise mathematical meaning of Taniyama's statement as I described above, and should also know what I did. I refer the reader to Sections A2 and A3 of the Appendix for some more details of relevant matters. The reader may ask why there were so many people who called the conjecture in various strange ways. I cannot answer that question except to say that many of them had no moral sense and most were incapable of having their own opinions.

Once, somebody asked me what I thought when I heard that an important case of my conjecture was verified. I responded, "I told you so," which was literally true.

## A2.  A Letter to Freydoon Shahidi

September 16, 1986

Dear Freydoon:

I now realize that I should have assumed that you were not familiar with what had been known in 1967. So let me explain about various things in more detail.

At a party given by a member of the Institute in 1962-64, Serre came to me and said that my results on modular curves (see below) were not so good since they didn't apply to an arbitrary elliptic curve over $\mathbf{Q}$. I responded by saying that I believed such a curve should always be a quotient of the jacobian of a modular curve. Serre mentioned this to Weil who was not there. After a few days, Weil asked me whether I really made that statement. I said, "Yes, don't you think it plausible?" Then he said, "Since both sets are countable, ... ... " as he translates it in French on page 450 of his complete work, Vol. III.

On page 454, he is sort of against the idea of making conjecture. For this reason, I think, he avoided to say in a straightforward way that I stated the conjecture.

As for the modular and other curves, let me mention Eichler's paper in 1954 and my three papers: J. Math. Soc. Japan vol. 10 (1958), 1–28; vol. 13 (1961), 275–331; Ann. of Math. 85 (1967), 58–159.

In these papers, it is shown that the zeta function of an "arithmetic quotient (especially a modular) curve" has analytic continuation. The same applies to its jacobian. The result about its quotient is not explicitly mentioned, but it is an easy consequence of the fact that the Hecke operators, as algebraic correspondences, are defined over $\mathbf{Q}$ or an appropriate number field.

I was conscious of this fact when I talked with Serre. In fact, I explained about it to Weil, perhaps in 1965. He mentions it at the end of his paper [1967a]: "nach eine Mitteilung von G. Shimura ... ." I even told him at that time that the zeta function of the curve $C'$ mentioned there is the Mellin transform of the cusp form in question, but he spared that statement. Eventually I published a more general result in my paper in J. Math. Soc. Japan 25 (1973), as well as in my book (Theorems 7.14 and 7.15).

Of course Weil made a contribution to this subject on his own, but he is not responsible for the result on the zeta function of modular elliptic curves, nor for the basic idea that such curves will exhaust all elliptic curves over $\mathbf{Q}$.

If you have any further questions, please let me know.

Sincerely yours,
Goro Shimura

## A3. Two Letters to Richard Taylor

November 25, 1994

Dear Richard:

I had the pleasure of reading your article "On the $\ell$-adic cohomology of Siegel threefolds," Invent. math. 114 (1993), 289–310, and got the impression that you didn't know the following paper of mine:

Construction of class fields and zeta functions of algebraic curves, Ann. of Math. 85 (1967), 58–159.

I would like to call your attention to the fact that in the last section of this paper the zeta and L-functions of an algebraic curve $V$ are determined, where $V$ is obtained from a quaternion algebra over a totally real algebraic number field unramified at only one archimedean prime.

I rarely (almost never) write a letter of this type, but in the present case I have a good reason for doing so. In fact, strangely this paper has not been mentioned in all the review articles and even in the original research papers. At best the authors mention that I treated the case of quaternion algebras over $\mathbf{Q}$, which is true, but that information alone is misleading. Clozel's Bourbaki article (March 1993, No.766) is an example. Once somebody writes a review article, then other people simply follow it without making any historical research on their own. Consequently the misconception is perpetuated, which is why I am writing.

Therefore may I ask you to read at least the introduction and the relevant part of the above 1967 paper? I wonder if you have any interest in the historical aspect of the matter. Assuming that you do, I am allowing myself to enclose a copy of my letter to Ken Ribet and hope that you will find it interesting. In it I did not explain everything. For example, the trace formula was very much relevant to my ideas behind my conjecture that every $\mathbf{Q}$-rational elliptic curve is modular. There is another strange fact: nobody has asked me why I made the conjecture. Perhaps it looks so obvious now.

Sincerely yours,

Goro Shimura

December 12, 1994

Dear Richard:

It is as if I forced you to ask that question. Anyway here is what I can say about my ideas which led me to the conjecture that every **Q**-rational elliptic curve is modular.

First of all, there is a basic philosophy that if a certain property of a mathematical object (the zeta function of a **Q**-rational elliptic curve) is best described in a special presentation of that object (uniformization by modular functions), then it is natural to expect that all the objects of the same type must be presented in the same manner. Of course this may turn out to be wrong, but one can always take it as a starting point. However, in addition to such a crude philosophy, there are at least two technical facts which support the idea.

The first of these is relatively simple. I already wrote about this in a letter to Andrew, and so let me quote a passage from it:

I saw Bryan Birch at a summer conference at Boulder in 1963. He told me his ideas on the significance of $Z_E(1)$, where $Z_E$ is the zeta function of a **Q**-rational elliptic curve $E$. He knew nothing about the curves uniformized by modular functions. So I explained to him the results of Eichler and myself. Furthermore, since he was naturally interested in the question of vanishing or nonvanishing of $Z_E(1)$, I told him the following three facts:

(1) If $E$ is given as $\Gamma_0(N)\backslash H$, then $Z_E(1)$ is a constant times $\int_0^\infty f(iy)dy$ with the cusp form $f$ in question, and therefore $Z_E(1)$ is practically a period. If $N = 11$ for example, the explicit form of $f$ tells that $f(iy)$ is always positive, so that $Z_E(1) \neq 0$, and the same holds in some other cases.

(2) If $g$ is the twist of $f$ by a Dirichlet character, then $g$ is a cusp form of higher level.

(3) In particular, if the character is quadratic, then the Mellin transform of $g$ gives $Z_D$ for the twist $D$ of $E$. From the functional equation of $Z_D$ we can obtain examples of $D$ such that $Z_D(1) = 0$.

These, if easy, were all new to him. He acknowledges this in his papers in the early 1960's. I think I told him only about the cases in which $\Gamma_0(N)\backslash H$ is of genus 1. Did I tell him that $D$ was a factor of the jacobian at the higher level? I think I knew it, but probably I didn't tell him about it beyond (2) and (3). (End of quotation)

In short, if $E$ is modular, its twist is modular too.

Now the second fact is a stronger factor. As you know I was working on the varieties of moduli of abelian varieties, since for me that was the only way to get hold of the zeta function of a variety. I first studied the case of the family of abelian varieties whose endomorphism algebras contain an indefinite quaternion algebra $B$ over $\mathbf{Q}$ (Proc. Int. Cong. M. 1958, J. Math. Soc. Japan, 1961) and found good curves with good Euler products. I started also the investigation of more general cases of quaternion algebras over a totally real number field $F$. However, if $F \neq \mathbf{Q}$, Hecke theory for such an algebra produces Euler products over $F$, and besides, in the case in which we get algebraic curves, the natural field of definition is $F$ or its abelian extension (Ann. of Math. 76 (1962), Osaka Math. J. 14 (1962)). Therefore, to get a $\mathbf{Q}$-rational object, I had to take $F$ to be $\mathbf{Q}$. At first I thought that these curves obtained from a division $B$ over $\mathbf{Q}$ might not be modular, (and strictly speaking that is true, see below), but I realized that *no nonmodular $\mathbf{Q}$-rational elliptic curves could be obtained* for the following reason: Eichler showed by means of his trace formula that the Euler products on $B$ are already included in those obtained from elliptic modular forms (Acta Arith. 4 (1958)). This result was later generalized by Shimizu (Annals 81 (1965)). The so-called Tate conjecture was explicitly stated much later, but the idea was known to Taniyama and myself, and so it was natural for me to think that two elliptic curves with the same zeta function are isogenous.

This latter fact concerning $B$ may have been the strongest reason for my stating the conjecture. Taniyama said "other special types of automorphic functions are necessary," and he meant Hecke's nonmodular triangle functions, but I never thought they were necessary. (Well, he may turn out to be right.)

As to the curves obtained from division $B$, I showed that the natural models of the curves have *no real points* (Math. Ann. 215 (1975)) even when the genus is 1, and in that sense they are not modular. They are not *elliptic curves*, though their jacobian varieties are. This point may explain the raison d'être of those curves. I wonder if there is any recent investigation on this phenomenon.

Finally let me note an episode in connection with the above problem. In the summer of 1962 Ihara told me (when we were in a Tokyo coffee shop, I think) that he had found an equality between two trace formulas. I was aware of the above 1958-paper of Eichler, and so we both went to the University Library to check it and found that Ihara's result was included in Eichler's. I guess he was disappointed, but I remember nothing in that respect. At that time he was a graduate student at the University of Tokyo, and I came to Princeton in September that year. He eventually wrote his Master's thesis on this topic, but never published it.

I can recall some more episodes in that period concerning the trace formula and related subjects, but I will tell you about them some other time, perhaps when you come to Princeton.

Sincerely yours,

Goro Shimura

*Notes added in 2007:*

1. In the last letter I wrote, "Well, he may turn out to be right." This is because the conjecture had not been completely settled by 1994.

2. The reader is also referred to *Notes* to articles [64e] and [89a] in my *Collected Papers,* vols. I, IV, which include some more explanations about what I was thinking or doing in the 1950s and 1960s.

## A4. Response

Notices of the American
Mathematical Society
vol. 43, No. 11 (November 1996), 1344–1347

(This is my response when I received Leroy P. Steele Prize for Lifetime Achievement from the American Mathematical Society at the Summer Mathfest held at the University of Washington, August 1996.)

I always thought this prize was for an old person, certainly someone older than I, and so it was a surprise to me, if a pleasant one, to learn that I was chosen as a recipient. Though I am not so young, I am not so old either, and besides, I have been successful in making every newly appointed junior member of my department think that I was also a fellow new appointee. This time I failed, and I should be grateful to the selection committee for discovering that I am a person at least old enough to have his lifetime work spoken of.

There are many prizes conferred by various kinds of institutions, but in the present case, I view it as something from my friends, which makes me really happy. So let me just say thank you, my friends!

*           *           *

I would like to take this opportunity to give a historical perspective of a topic on which I worked in the 1950s and 1960s, intermingled with some of my personal recollections. It concerns arithmetic Fuchsian groups which can be obtained from an indefinite quaternion algebra $B$ over a totally real algebraic number field $F$. For such a $B$ one has

$$B \otimes_{\mathbf{Q}} \mathbf{R} = M_2(\mathbf{R})^r \times \mathbf{H}^{d-r},$$

where $d = [F : \mathbf{Q}]$, $0 \le r \le d$, $M_2(\mathbf{R})$ is the matrix algebra over $\mathbf{R}$ of size 2, and $\mathbf{H}$ is the Hamilton quaternions. Assuming $r > 0$ and taking a subring $R$ of $B$ that contains $\mathbf{Z}$ and spans $B$ over $\mathbf{Q}$, denote by $\Gamma$ the group of invertible elements of $R$ whose projection to any factor $M_2(\mathbf{R})$ has determinant 1. Then we can view $\Gamma$ as a subgroup of $SL_2(\mathbf{R})^r$ through the projection map to $M_2(\mathbf{R})^r$, and so we can let $\Gamma$ act on the product $H^r$ of $r$ copies of the upper half plane $H$. In this way we obtain an algebraic variety $\Gamma \backslash H^r$, which is an algebraic curve if $r = 1$. It is known that $\Gamma \backslash H^r$ is compact if and only if $B$ is a division algebra. In particular, we can take $B$ to be the matrix algebra $M_2(F)$ over $F$ of size 2, in which case $r = d$ and the meromorphic functions on $\Gamma \backslash H^d$ are called *Hilbert modular functions*.

If $F = \mathbf{Q}$, the group $\Gamma$ was first discovered by Poincaré [7], apparently in 1886. He reminisced in his *Science et Méthode* as follows: "One day, while walking on a cliff, it occurred to me, with the same customary characters of shortness, suddenness, and immediate certainty, that arithmetic transformations of indefinite ternary quadratic forms were identical to those of non-Euclidean geometry."

One interesting aspect of this work is that the quotient $\Gamma \backslash H$ is compact if $B$ is a division algebra. Until then the only Fuchsian groups he or anybody else knew were those obtained from hypergeometric series, among which the arithmetically defined ones were the classical modular groups; in all those cases the quotient is not compact. (Uniformization of an arbitrary compact Riemann surface was proved independently by Koebe and Poincaré only in 1907.) Poincaré's group was generalized to the case $1 = r \le d$ with an arbitrary $F$ by Fricke [3] in 1893. It is also discussed in the last chapter of the thick volume [4] of Fricke and Klein published in 1897. These mathematicians employed an indefinite ternary quadractic form instead of a quaternion algebra. Since $SO(2, 1)$ is covered by $SL_2(\mathbf{R})$, the unit group of the given ternary form produces a discrete subgroup of $SL_2(\mathbf{R})$.

After Fricke's investigations, which showed that the action of the groups on $H$ is properly discontinuous, no significant progress was made in this area for the next fifty years. In 1912, Hecke published his thesis work [5] concerning Hilbert modular functions in the case of $M_2(F)$ with $d = 2$. In its introduction he said that the results of Fricke on the Fuchsian groups of the above type seemed to be "without specific meaning in number theory." Later developments proved that he was wrong. Taking his tender age of twenty-five into consideration, we may forgive him, and may even justify his comment, allowing him a thirty-year warranty, since it could apply to all papers on this subject in that period, one by Heegner [6], for example, which I cite here in order to show that the topic was not forgotten but was being treated without any new ideas. It should also be pointed out that Hecke's own work was critically flawed, though generally speaking he was headed in the right direction, except for that comment.

Eichler may have been the first person who was seriously interested in this group. He wrote his dissertation with Brandt on quaternion algebras, and later worked on more general types of simple algebras. He once told me that Brandt did not think much of non-quaternion algebras, and was unhappy with Eichler's turning to them. In reality, there was no need for him to be unhappy, since the fact that Eichler started with quaternion algebras determined his course thereafter, which was vastly successful. In a lecture he gave in Tokyo, he drew a hexagon on the blackboard, and called its vertices clockwise as follows: automorphic forms, modular forms, quadratic forms, quaternion algebras, Riemann surfaces, and algebraic functions. Anyway in the mid-1950s Eichler was developing the theory of Hecke operators for the Fuchsian groups of Poincaré's type (see [1], for example). He also gave a formula for the genus of $\Gamma \backslash H$ somewhat earlier. However, there were no other number-theoretical investigations on these algebraic curves by that time.

In 1957 while in Paris I became interested in this class of groups. I had just finished my first work on the zeta functions of elliptic modular curves. Though I knew that it needed elaboration, I was more interested in finding other curves whose zeta functions could be determined. I was also trying to formulate the theory of complex multiplication in higher dimension in terms of the values of automorphic functions of several variables, Siegel modular functions, for example. It turned out that these two problems were inseparably connected to each other. Also, nobody else was working on such questions. I can assure the reader that I had no intention of humiliating Hecke posthumously.

So I took up the group of the above type. My aim was to find an algebraic curve $C$ defined over an algebraic number field $k$ that is complex analytically isomorphic to $\Gamma \backslash H$, and to determine the zeta function of $C$. Such a $C$ is called a *model of* $\Gamma \backslash H$ *over* $k$. Naturally I started with the simplest case $F = \mathbf{Q}$. Since it was relatively easy to see that $\Gamma \backslash H$ in this case parametrizes a family of certain two-dimensional abelian varieties, I was soon able to prove that the curve had a $\mathbf{Q}$-rational model. The proof required a theory of the field of moduli of a polarized abelian variety, but luckily, I had it at my disposal, since I had been forced to develop such a theory in order to get a better formulation of complex multiplication as mentioned above.

In June 1958 I visited three schools in Germany: Münster, Göttingen, and Marburg. I gave a talk at each place, but remember only that at Göttingen I spoke about the field of moduli of a polarized abelian variety, and its application to the field of definition for the field of automorphic functions. At the end of my talk I mentioned briefly the $\mathbf{Q}$-rationality of the curve $\Gamma \backslash H$ for Poincaré's $\Gamma$.

Siegel was among the audience, and pressed on the last point. I began to explain the idea, but he interrupted me and simply wanted to know whether I really had the proof. So I said "Yes," and that was that. Siegel said nothing, but apparently he was not convinced, and expressed his doubts to Klingen, who in 1970 told me about Siegel's skepticism at that time. I can easily guess the rationale behind his disbelief: since $\Gamma\backslash H$ is compact, there is no natural Fourier expansion of an automorphic form, so that there is no way of defining the rationality of automorphic functions, and that was exactly why Hecke made the comment mentioned above. Eventually I determined the zeta function of the curve, and gave a talk on that topic at the ICM, Edinburgh, in September 1958. The full details were published in [8] in 1961.

There was no such incident at Marburg, where I met Eichler. I remember that after dinner at his home, he played a religious piece of music on the phonograph, which I think was by Bach. I am sure it was not by Mozart, as he did not think much of the composer. He was a tall and handsome man, whose look immediately reminded me of the knight in the movie "The Seventh Seal" by Ingmar Bergman, which I had seen in Paris a few months earlier. As for Siegel, who was sixty-one at that time, calling him a big mass of flesh would have been misleading and even derogatory, but that was my first impression. Though he must have looked awesome to many, he assumed no airs, and there was a certain homely atmosphere around him, which made him less intimidating, at least to me.

Coming back to my work, at first I thought that these curves obtained from a division quaternion algebra $B$ over $\mathbf{Q}$ might not be modular, (and strictly speaking, that is true, see the next paragraph), but I realized that *no nonmodular* $\mathbf{Q}$-*rational elliptic curves could be obtained* for the following reason: Eichler had shown, by means of his trace formula, the following fact: the Euler products on $B$ are already included in those obtained from elliptic modular forms [2]. The Tate conjecture on this was explicitly stated much later, but the idea was known to many people, and so it was natural for me to think that two elliptic curves with the same zeta function are isogenous. This fact concerning $B$, in addition to the results I had about the zeta functions of modular curves, may have been the strongest reason for my stating the conjecture that every $\mathbf{Q}$-rational elliptic curve is modular.

Let me insert here a remark on the curves obtained from a division algebra $B$. I showed much later in [11] that the natural models of the curves have *no real points* even when the genus of $\Gamma \backslash H$ is 1, and in that sense they are not modular! They are not *elliptic curves* in the strict sense, though their jacobian varieties are. This point may explain the raison d'être of those curves. I wonder if there is any recent investigation on this phenomenon.

The curves with $F \neq \mathbf{Q}$ were more difficult. After going back to Tokyo in the spring of 1959, I decided to investigate more general families of abelian varieties. By specifying the types of endomorphism algebra and polarization of abelian varieties, one obtains a quotient $\Delta \backslash S$ that parametrizes abelian varieties of a prescribed type, where $S$ is a hermitian symmetric domain of noncompact type, and $\Delta$ is an arithmetic subgroup of a certain algebraic group. The above $\Gamma \backslash H$ for Poincaré's $\Gamma$ is an easiest example of $\Delta \backslash S$; one simply takes $B$ to be the endomorphism algebra. For a certain reason, however, the algebra $B$ with $0 < r < d$ never appears as the endomorphism algebra of an abelian variety, which was the main difficulty. Then I realized that choosing an algebra different from $B$, one obtains $\Delta \backslash S$ that is essentially the same as $\Gamma \backslash H$ for an arbitrary $B$ of the above type. I think that was sometime in the fall of 1960. I knew at that point that the problem was approachable, and even knew that the curves had models over a number field, but did not know how to state the theorems in the best possible forms, not to mention how to prove them.

In a series of papers published in 1963–65 I investigated the number fields over which the varieties $\Delta \backslash S$ can be defined. In many higher-dimensional cases, the results were best possible, but in the one-dimensional case that was the main question, I was not satisfied. So I turned to a higher-dimensional case of a different nature. In a famous paper on symplectic geometry [12] Siegel defined a certain arithmetic subgroup $\Gamma'$ of $Sp(n, \mathbf{R})$ which was a

generalization of Fricke's group, and which was also defined relative
to $F$. If $n > 1$ and $F \neq \mathbf{Q}$, this group does not appear as the above
group $\Delta$ associated with a family of abelian varieties. But in the
summer of 1963, while in Boulder, Colorado, I found that there was
an injection $\Gamma' \to \Delta$ with some $\Delta$, which produced a holomorphic
embedding $\Gamma'\backslash S' \to \Delta\backslash S$, where $S'$ is the Siegel upper half space
of degree $n$. If $n = 1$, $\Gamma'\backslash S'$ is exactly the algebraic curve $\Gamma\backslash H$
in question, and moreover the embedding is essentially birational
over $\mathbf{C}$. Anyway, employing this embedding, I was able to find a
number field over which $\Gamma'\backslash S'$ is defined for an arbitrary $n$. When
I was asked to contribute a paper to the volume in honor of Siegel's
seventieth birthday, I naturally took this as the topic, and sent the
manuscript to the editor in the fall of 1965.

Around the same time, perhaps in early September that year,
I finally had a definite idea of settling the original question in the
one-dimensional case: to employ many different $\Delta\backslash S$ for a given
$\Gamma\backslash H$. By means of this idea together with a finer theory of variety
of moduli of polarized abelian varieties, by June 1966 I was able to
finish the paper [9] in which I determined the zeta function of the
curve $\Gamma\backslash H$ with any totally real $F$. At the same time I determined
the class fields generated by the values of automorphic functions,
not only in the one-dimensional case but also in the case where $B$
is totally indefinite, including the Hilbert modular case. By doing
so, I showed that similar theories could be developed in a parallel
way in both Fricke's and Hecke's cases. In fact, those are the two
extreme cases of a more general class of arithmetic quotients for
which one can do number-theoretical investigations Hecke wished
to do in his case, a fact Hecke never realized.

I dedicated the paper to Weil. At some point I said to him
jokingly that he became sufficiently old that I could now dedicate a
paper to him, to which he replied, "I can't stop it." Meanwhile my
paper dedicated to Siegel appeared in the *Mathematische Annalen*
[10]; I also sent a reprint of my Annals article to him, as I had been

doing regularly with my earlier papers. Here is what he wrote me about these:

Göttingen, 15 May 1967

Dear Professor Shimura:

After a long trip around the world I returned to Göttingen and I found your last paper from the Annals of Mathematics together with the work which you kindly dedicated on the occasion of my 70th birthday.

I am sending you my most cordial thanks for your kindness. I have now begun to study these two papers, and both of them seem to be of great interest, from the arithmetical and the analytical point of view.

During many years I have regretted that Hecke's earlier work on Hilbert's modular function and class field theory had not been continued by later mathematicians. I am glad to see in your last paper how much you have already achieved in this direction.

I was very pleased to see from your other paper that you have obtained decisive results concerning those groups which I introduced in my paper on symplectic geometry.

Best congratulations for the success of your previous work, and best wishes for the future!

Yours sincerely
Carl Ludwig Siegel

I was naturally gratified and even moved, but frankly I was somewhat disappointed by his mentioning only the Hilbert modular case, which was far easier than the case of curves that was the main feature of my paper. Therefore I was not sure whether he perceived the full scope of the work. Perhaps he thought what he said was enough, which is true, and so I should not complain. In fact, reading this letter after almost thirty years, I now think that the letter tells more about the sender than about the recipient.

To clarify this point, we have to know what kind of a man Siegel was. Of course he established himself as one of the giants in the history of mathematics long ago. He was not known, however, for his good-naturedness. Around 1980 I sat next to Natasha Brunswick at a dinner table, when she proclaimed, "Siegel is mean!" I don't remember how our conversation led to that statement, but many of those who knew him would agree with her opinion. Hel Braun, one of his few students, apparently disliked him. He was indisputably original, and even original in his perverseness. Once at a party he played a piano piece and challenged the audience to tell who the composer was. Hearing no answer, he said it was a sonata by Mozart, Köchel number such and such, played backward. On the other hand, he had a certain sense of humor. When Weil asked him which work of his he thought best, he replied, "Oh, I think a few watercolors I made in Greece some years ago are pretty good."

In any case, it would be wrong to presume him to be a mathematician who did what he wanted to do, unconcerned about what other people might think of his work. I believe he was not that aloof. He must have known who he was, but at the same time he must have felt unappreciated by the younger generation. That was Eichler's opinion, and I am inclined to agree with him.

After his retirement, Siegel took a long trip around the world as he mentions in his letter. On coming back to Göttingen, one day he went into his office in the university, and found on his desk a copy of the volume of the *Mathematische Annalen* dedicated to him, which pleased him greatly. And here was a man thirty-four years younger than he, completely outside of his German influence, who took up the topic on which he expended considerable effort many years ago, with genuine appreciation of his work.

Perhaps he was not so crabbed as many people had imagined, and it is possible that he wrote a few more letters like the above one. At any rate, when he wrote that letter, he knew that at least one of his papers was really understood, and at that moment he was capable of appreciating the progress made by the new generation, of which he had often been contemptuous. I am indeed glad to be the recipient of the letter which showed this great mathematician as a warm-hearted man with no trace of ill-temperedness, nor any cynicism.

$$* \qquad * \qquad *$$

Here are some facts of supplementary nature to the above article. The first paragraph was already included in the Notes at the end of my *Collected Papers,* vol. IV. The second paragraph is added in January, 2008.

I mentioned a "lecture" by Eichler in Tokyo and "a hexagon" he drew on the blackboard. Actually he gave several lectures from April 12 through 24, 1958 at the University of Tokyo. Notes were taken by Y. Taniyama, and published (in Japanese) in Sugaku 10 (1959), 182–190. I did not attend the lectures, as I was in Paris at that time, but found the hexagon in the notes. There was a colloquium in honor of Eichler's seventieth birthday at the University of Basel, June 4 and 5, 1982, and I was asked to talk on his work, which I gladly did. At the beginning of the talk I drew on the blackboard a hexagon whose vertices were named as in the present article, saying that it was the same as what he drew twenty-four years ago. He was much amused by it and said that he had completely forgotten about it.

I also mentioned Hel Braun in connection with Siegel. According to Chowla, he would often fetch her for a walk, and tell her that if she ever were to marry, her mate must be a first-rate mathematician. She eventually became the companion of Emil Artin during his late years at Hamburg. He had moved there from Princeton in 1958 and died in 1962. After my talk at the Moscow Congress in 1966, Braun approached me and offered some compliments. This incident firmly registered in my mind, as it happened very rarely in my career. Possibly I was better known in those days than now, and I have been slipping since then.

# References

[1] M. Eichler, Modular correspondences and their representations, J. Ind. Math. Soc. **20** (1956), 163–206.

[2] M. Eichler, Quadratische Formen und Modulfunktionen, Acta arith. **4** (1958), 217–239.

[3] R. Fricke, Zur gruppentheoretischen Grundlegung der automorphen Functionen, Math. Ann. **42** (1893), 564–597.

[4] R. Fricke and F. Klein, *Vorlesungen über die Theorie der automorphen Funktionen,* I. Leipzig, Teubner, 1897.

[5] E. Hecke, Höhere Modulfunktionen und ihre Anwendung auf die Zahlentheorie, Math. Ann. **71** (1912), 1–37 (=Werke, 21–57).

[6] K. Heegner, Transformierbare automorphe Funktionen und quadratishe Formen I, II, Math. Z. **43** (1937), 162–204, 321–352.

[7] H. Poincaré, Les fonctions fuchsiennes et l'arithmétique, J. de Math. 4 ser. 3 (1887), 405–464 (=Oeuvres, vol. 2, 463–511).

[8] G. Shimura, On the zeta functions of the algebraic curves uniformized by certain automorphic functions, J. Math. Soc. Japan, **13** (1961), 275–331.

[9] ———, Construction of class fields and zeta functions of algebraic curves, Ann. of Math. **85** (1967), 58–159.

[10] ———, Discontinuous groups and abelian varieties, Math. Ann. **168** (1967), 171–199.

[11] ———, On the real points of an arithmetic quotient of a bounded symmetric domain, Math. Ann. **215** (1975), 135–164.

[12] C. L. Siegel, Symplectic Geometry, Amer. J. Math. **65** (1943), 1–86(=Gesammelte Abhandlungen, II, 274–359).

## A5. André Weil as I Knew Him

Notices of the American Mathematical Society
vol. 46, No. 4 (April, 1999), 428–433

Bathed in the sunlight of late summer, I was walking a quiet street of Takanawa, a relatively fashionable district in southern Tokyo, toward the Prince Hotel Annex, where André Weil was staying. It was the afternoon on a warm day of early September in 1955. He was among the nine foreign participants of the International Symposium on Algebraic Number Theory, to be held in Tokyo and Nikko that month. The Korean war had ended two years earlier, and in the United States, Eisenhower's first term had begun in the same year. Five years later, in 1960, his planned visit to Japan would be hindered by the almost riotous demonstrations of labor unions and students in the city, but nobody foresaw it in the peaceful atmosphere of the mid 1950s. While walking, I had a mildly uplifted feeling of expectation and curiosity about what would happen, the first of those I would experience many times later, whenever I was going to see Weil.

My acquaintance with him began in 1953, when I sent my manuscript on "Reduction of algebraic varieties with respect to a discrete valuation of the basic field" to him in Chicago, asking his opinion. I told him my intention of applying the theory eventually to complex multiplication of abelian varieties. In his answer, dated December 23, 1953, he was quite favorable to the work, and encouraged me to proceed in that direction; he also advised me to send the paper to the *American Journal of Mathematics*, which I did. By that time I had read his trilogy *Foundations, Courbes algébriques,* and *Variétés abéliennes,* as well as his 1950 Congress lecture [50b][1] and a few more papers of his. I was also aware of the existence of many of his other papers, or had some vague ideas

about them, [28], [35b]<sup>2)</sup>, [49b], [51a], for example. But I don't
think I had read all those before 1955. The article "L'avenir des
mathématiques" [47a] and his review [51c] of Chevalley's book on
algebraic functions were topics of conversation among young mathematicians in Tokyo. Later, while in Japan, when he was asked
to offer his opinion on various things, he jokingly complained that
he was being treated like a prophet, not a professor. But to some
extent that was so even before his arrival.

In any case, when he accepted the invitation to the Tokyo-
Nikko conference, we young mathematicians in Japan expected
him with a sense of keen anticipation. I shook hands with him for
the first time on August 18, in a room in the Mathematics Department, University of Tokyo. He looked gentler than the photo
I had seen somewhere. He was forty-nine at that time. Our meeting was short, and there was not much mathematical discussion,
nor did he make any strong impression on me that day. He was
given, perhaps a few days later, a set of mimeographed preliminary
drafts of papers of most Japanese participants, including my 49-
page manuscript titled "On complex multiplications", which was
never published in its original form.<sup>3)</sup>

About two weeks later a message was forwarded to me: Professor Weil wishes to see me at his hotel. So I brought myself
there at the appointed time. He appeared in the lobby wearing
beige trousers with no jacket, or tie. He had read my manuscript
by then, and sitting on a patio chair in a small courtyard of the
hotel, he asked many questions and made some comments. Then
he started to talk about his ideas on polarization of an abelian
variety and a Kummer variety. He scribbled various formulas on
some hotel stationery, which I still keep in my possession. At some
point he left his chair; pacing the courtyard from one end to the
other, he impatiently tried to pour his ideas into my head. He
treated me as if I was an expert who knew everything. I knew of
course what a divisor meant and even the notion of linear and algebraic equivalence, as I had read his 1954 *Annalen* paper on that

topic, but I lacked the true feeling of the matter, not to speak of the historical perspective. Therefore, though I tried hard to follow him, it is fair to say that I understood little of what he said. At the end we had tea, and he ate a rather large piece of cake, but I declined his offer of the same, perhaps because his grilling lessened my appetite.

During the conference and his stay in Tokyo afterward, I saw him many times. On each occasion he behaved very naturally, if in a stimulating way. It was as if that hotel encounter had the effect of immunization for me, and possibly for him too. I remember that I asked him about the nature of the periods of a differential form of the first kind on an abelian variety with complex multiplication. He said, "They are highly transcendental," which was not a satisfactory answer, but as good as anything under the circumstances. At least, and at long last, I found someone to whom I could ask such a question. Those several weeks were truly a memorable and exciting period. To make it more exciting, one of my colleagues would make a telephone call to the other, imitating Weil's voice and accent: "Hello, this is Weil. I didn't understand what you said the other day; so I'd like to discuss with you ... ." Sometimes the prank worked. A few days before his leaving Tokyo for Chicago, I, together with three such naughty boys, visited him in another hotel in the same area. Taniyama promised to come, but didn't; apparently he overslept as usual. During our conversation, Weil advised us not to stick to a wrong idea too long. "At some point you must be able to tell whether your idea is right or wrong; then you must have the guts to throw away your wrong idea."

As he said in his *Collected Papers*, his stay in Japan was one of his most enjoyable and gratifying periods. He found an audience of young people who were not afraid of him and was sophisticated enough to understand, or at least willing enough to try to understand, his mathematics; he certainly had an audience in the United States then, but apparently of a different kind.

More than two years passed before I saw him again, which was in Paris in November 1957. Henri Cartan, accepting his suggestion, had secured a position of chargé de recherches at CNRS for me. Weil was on leave from Chicago for one year, and sharing an office with Roger Godement at the Institut Henri Poincaré, but he occupied it alone for most of the time. In that period he was working on various problems on algebraic groups, the topics which can be seen from [57c], [58d], and [60b], for example. He was giving lectures on one such subject at the École Normale Supérieure and regularly attended the Cartan seminar. He lived in an apartment at the southeast corner of the Luxembourg garden with a fantastic view of Sacré Coeur to the far north and the Eiffel tower to the west. One of his favorite restaurants was Au vieux Paris, in the back of the Panthéon. A few days after my arrival, he invited me to have lunch there. I remember that he had radis au beurre (radish with butter) and lapin (rabbit) sauté, a fairly common affair in those days, but perhaps somewhat old-fashioned nowadays. I don't remember his choice of wine, but most likely a fullsized glass of red wine for each of us. To tell the truth, it was not rare to find him snoozing during the seminar. From his apartment, the institute and the Panthéon could be reached in less than ten minutes on foot. Paris in the 1950s retained its legendary charm of an old city, which had not changed much — he once told me — since the days of his childhood. It is sad to note that the city went through an inevitable and drastic transformation in the 1970s.

Though I was working on a topic different from his, he was earnestly interested in my progress, and so I would drop into his office whenever I had something to talk about. For instance, one day I showed him some of my latest results for which I employed Poincaré's theorem on the number of common zeros of theta functions. He smiled and said, "Oh, you use it, but it is not a rigorously proved theorem." Then he advised me to take a different route, or to find a better proof; later he told me a recently proved result concerning divisors on an abelian variety, by which I was able to save my result, as well as Poincaré's theorem.

On another occasion, I heard some shouting in his office. As I had only a brief message to him, I knocked on the door. He opened it and introduced me to Friederich Mautner[4], professor at Johns Hopkins, who was his shouting partner. After a minute or so I left. As soon as I closed the door, they started their shouting again. When I was walking through the corridor after spending half an hour in the library, the shouting match was still going on; I never knew when and how it began and ended, nor who won.

From time to time he fetched me for a walk in the city. The topics of our conversation during those walks were varied; he would suggest to me, for example, that I go to churches to listen to religious music; he said it was necessary for me only to stand up and sit down when others did. When asked about his faith, he said, "Pas du tout" ("Not at all"). According to him, one of the best way to learn French or any foreign language was to see the same movie in that language again and again, staying in the same seat in the same movie theatre, a piece of advice I followed perhaps too faithfully. It was the time when Brigitte Bardot and Zizi Jeanmaire were at their zenith. Another method he suggested was to read newspapers, but I was not so dilligent in this task. Perhaps as he became impatient with my slow progress in French, he asked me whether I was doing my homework in that respect. I dodged the issue by mentioning an old Oriental saying, "He who runs after two rabbits will catch neither." Maybe I subconsciously remembered the rabbit for his meal. "What's your rabbit? Hecke operators?" he asked. Then we discussed about the possible method or philosophy of how a Frobenius of a reduced variety can be lifted. A few days later he caught me in the library and asked again, "What about your rabbit?" He was an extremely sharp man, and clearly he sensed that I was up to something, which was true. In this article, however, I should leave the rabbit at large, merely mentioning that he would later say, "How is your rabbitry doing?"

Starting in the fall of 1958 he was at the Institute for Advanced Study permanently, and I was a member there for that academic year. So I practically followed him, and I had the same daily routine with him for another several months. Looking back on those days, I am filled with a sense of deep gratitude to him for paying such an unusual and personal attention to me; also, I must note, to my regret, that unaware of the real meaning of my situation at that time, I did not take full advantage of my fortunate privilege of being constantly with such an extraordinary man in his prime.

In the spring of 1961, he spent a few months in Japan with his wife, Eveline. Though they undoubtedly enjoyed their stay, and I was happy to have a person at hand who really understood me — perhaps the only one at that time — I may be excused for saying that overall his presence was less than a pale revival of his former visit.[5] As for myself, after spending three years in Japan, I came back to Princeton in September 1962, when I began a new and long chapter of my relationship with him. To continue my narrative, I will now present some interesting aspects of his words and deeds in this period irrespective of the chronological order of the events.

As already mentioned above, he liked to walk, partly for the purpose of physical exercise. In Princeton every Sunday he would walk one and half miles from his home to buy the Sunday *New York Times*, and so, according to his daughters, his church denomination was pedestrian. At the Institute he would occasionally pick a walking partner among the members. He was not a good walker, however. Though he was physically fit and walked briskly, he often fell on his face by tripping on something on the ground. That happened when I was with him in the Institute wood, but I pretended to have seen nothing, as he hated being helped on such an occasion. Though he was not injured then, he was not so lucky other times. During such a walk, he would answer my questions, or would tell his stories. Here are some samples:

When he was twelve or thirteen,[6] there was a magazine for elementary mathematics asking the reader to send in solutions to the problems; then they would print best solutions. He contributed many as he found great pleasure in seeing his name in the magazine, but he graduated from that level after about two years. Then he said: "Maybe I should have included some of the solutions in my oeuvres, he! he! he!"

Around the time when he was at Haverford, he asked Hermann Weyl to lend him some money. "How much?" asked Weyl. "Well, four or five hundred dollars." Then Weyl brought out his checkbook, and after thinking awhile he signed a check for four hundred and fifty dollars.

When he was teaching at Lehigh, a student asked him for help in calculus. After they spent a lot of time struggling to find out what his problem was, the student finally said, "I don't seem to understand this symbol $x$."[7] He referred to his Lehigh days as his period of "overemployment".

A French gentleman's ideal is to have three concurrent loves: the first one, whom he cares about at present; the second, a potential one, whom he has his eye on with the hope that she will eventually be his principal love; the third, the past one, with whom he hasn't completely cut off his relations. Then he observed: "It's a good idea for a mathematician to have three mathematical loves in the same sense."

He would talk about Baudelaire, Proust, and Gide, their homosexuality in particular, Paul Claudel's treatment of his sister Camille, and also about the letter exchange between Paul Claudel and Madeleine Gide. He amused himself by twisting each story in his own fashion to make it funny, often with a piquant effect.

I asked, for what reason I don't remember, whether he read detective stories. "Yes, but only when I have a cold," he said, and added, "You know, when you have a cold, there is nothing else to do but read detective stories." He was rather apologetic, and so I asked, "How often do you catch cold?" "Very often" was his answer.

As to Fields medals, he said: "It's a kind of lottery. There are so many eligible candidates, and the whole selection process is a matter of chance. Therefore the prize could be given to *any of them* as in a lottery."[8]

He used to say that a good mathematician must have two good ideas. "It is possible for someone to have a really good idea, but it may be just a fluke. Once the person has a second good idea, then there is a good chance for him to develop into a better mathematician." He mentioned a well-known American as a prolific mathematician with a single idea. He also noted Mordell as a counterexample to his principle.

He could say something even harsher, but that was rare. In the summer of 1970 after the Nice Congress, I was talking with him somewhere in the Institute about French mathematicians. He observed that there were three young mathematicians in Paris who started brilliantly, and so there were high expectations for them. He mentioned three well-known names and said, "What happened to them? They utterly failed to produce anything great." That was more than a quarter century ago, but I cannot tell whether or not he changed his opinion, as we never talked on that matter again. After around 1975 he expressed, more than once, his pessimistic view that French mathematics had been declining for some time. Therefore we should perhaps take his criticism in that context.

He held Riemann and Poincaré in high esteem, which was more than natural; Hecke was also his favorite. He rarely talked about Hilbert in our conversation. He didn't think much of Klein, which is not surprising. Picard was depicted by him as formal and stiff. Among his contemporaries, he thought highly of Siegel, and spoke of Chevalley in amicable terms, but not so with Weyl, about whom he seemed to have a kind of ambivalence. He recognized the unusual talent of Eichler.[9] Hadamard was his teacher, and their relationship is well documented in his autobiography. He paid due

respect to Hasse, though he remembered the fact that Hasse wore a
Nazi uniform at some point.[10] He told me several anecdotes about
Hardy, but he presented each story in a sarcastic tone. "Hardy's
opinion that mathematics is a young man's game is nonsense," he
said.

It may be too optimistic a view to say that most people mellow
with advancing age. At least many do, and there are those who
don't. It is told, for example, that Saint-Saëns achieved an ever-
increasing reputation as a man of bad temper through his long
life of eighty-six years. Weil did mellow, but even after the age of
seventy he was capable, if rarely, of being childishly irritable, as can
be seen from the following episode. But first let me note: Around
1976 or 1977 he declared, "I am no longer a mathematician; I
am a mathematical historian." Apparently he realized that there
were no more subjects he could handle better than the younger
generation. Coming to my story: In my teens I somehow got
hold of a copy of a pirate edition, which was being called the
Shanghai edition, of *Eindeutige Analytische Funktionen* by Rolf
Nevanlinna. I enjoyed reading the first one-third of the book, but
gave up on the rest. Still, my reading of the book remains as one
of my fond memories. When I recognized Nevanlinna in a lecture
hall at the 1978 Helsinki Congress, I introduced myself and shook
hands with him, an incident which, in my youth I never imagined,
would happen. He was eighty-three then. Weil gave a lecture titled
"History of mathematics: Why and how" there.

After the congress, I spent a week in Paris, and one day I
was sipping coffee with Weil in a café near his apartment. I told
him about that happy experience of mine at Helsinki. But he
was much displeased with my story. He said with a grimace that
Nevanlinna was not such a good mathematician who was worthy
of my esteem, and so on. I was dumbfounded; I never idolized
Nevanlinna, whose name I knew before acquainting myself with
any of Weil's works, simply because the book was accidentally
available. That must have been clear to him. After all, it was

none other than Nevanlinna who saved him from being executed by the Finnish police, a fact he told me some years earlier, and narrated in his autobiography, which also includes a passage of the Weil couple's happy stay in Nevanlinna's villa in 1939.

I should add, however, that he could be found on the other side of the world. When there was a discussion of a new appointment at the Institute, Morton White, professor of the school of history, was fiercely against the proposition, and at the faculty meeting he expressed his opinion in a heated fashion. Then Weil, sitting next to him, said, "Calm down, please, calm down." White later told me that he thought the scene rather funny in view of the normal temperament of Weil.

After Eveline's passing away in May 1986 at the age of seventy-five, his daughter Nicolette bought a microwave oven for him. However, saying that he didn't like to "push the button," he never touched it, and so the oven was returned to the dealer. The Weils had been our regular dinner guests, but since then naturally he alone was with us, which happened not infrequently. It was sometime in December 1987. Weil, Hervé Jacquet, Karl Rubin, Alice Silverberg, my wife Chikako, and I had dinner at a Chinese restaurant, and were having dessert at our place. When I prodded the guests to tell their ambitions in their next lives, Jacquet said he would like to be an opera singer, and that was not a joke for him. In fact, opera singing was his first love, mathematics being merely the second. Next, "I want to be a Chinese scholar studying Chinese poems," said Weil. After visiting China twice, he had been reading English translations of Chinese standard literature like *The Dream of the Red Chamber*. "That may be a rather dull life, and I don't think a person like you can stand it," said I. "All right then, I will be a house cat. The life of a house cat is very comfortable." Pointing to our neighbor's female white cat who was also a guest, he said, "Maybe she will be my mother." Then Rubin said, "Perhaps a Chinese cat is a good solution." With laughter everybody accepted it. That was about a week or two before Christmas, and so after a few days, Chikako brought him a stuffed cat as a Christmas present, which pleased him greatly. In fact, the Weil family used to have a cat, and once he defended himself for having a Christmas tree in his house by saying that they had it because their cat loved it.

He was conscious of his old age, particularly after he became a widower. According to what he said: Eveline was afraid of becoming senile. But she was not at all senile when she died. A famous French mathematician, who lived beyond eighty, was senile in his last two years, but he knew it himself. So when he had visitors, he held a newspaper to show that he was at least able to read, but the paper was often upside down. Another, who lived longer, was not like that; even so, when Weil visited him, he brought out and showed him, one after another, the diploma of each of the many honorary degrees he had received.

As for Weil himself, he showed no such sign, as far as I remember. I talked with him sometime in November 1995 for half an hour or so in his office. He was alert and able to make a reasonable judgment on the matter for which I went to see him. There was a lunch party for his ninetieth birthday in May 1996 at a restaurant in Princeton; though he did not talk much, he was in a good mood. Before and after that Chikako had lunch at the Institute cafeteria several times; she would find him eating mostly alone, sometimes with his daughters; she would say hello to him, to which he would reply, "Is Goro here?" So she was relieved to find that he at least remembered her as someone related to me.

I saw him for the last time on December 19, 1996. For some reason he phoned me the day before. Since he had hearing difficulties, he finally suggested that I see him at the Institute. I proposed some date, but he said, "No, why don't you come tomorrow; otherwise I won't remember." So I had lunch with him there that day. From the previous night it had been drizzling endlessly. When I met him in the common room of Fuld Hall, he did not have his hearing aid, and he asked me to drive him home to get it. After getting it we went into the dining hall. He used to eat well, and almost twice as much as myself. Around 1980, André, Eveline, Chikako, and I had lunch together at a restaurant in New Hope, Pennsylvania. That was a buffet style affair, and he was in high spirits. I remember that his appetite impressed the remaining three. Incidentally, he was not fussy about wine. Not that he did not care, but it is my impression that Eveline cared more.

I was curious how he would eat this time. Not surprisingly, compared with what he ate sixteen years ago, the quantity he took was modest, less than half of the previous meal. Since he had hearing problems, it was difficult to conduct our conversation smoothly, and I often had to write words and sentences on a piece of paper. Unlike the occasion forty-one years ago, this time it was I who was writing. I was working on the Siegel mass formula[11] with a new idea at that time, and that was one of his favorite topics. So I asked him about the history of that subject. For example, I asked him whether or how he studied the works of Eisenstein, Minkowski, and Hardy. He said he didn't remember about Eisenstein,[12] but he had studied a little, but not much, of Minkowski's work; he never studied Hardy. He kept saying that it was a long time ago, and so he didn't remember, which must be true, and so we should not accept what he said at face value. In fact, to check that point, I asked him whether Minkowski was reliable. He said, "I think so." At that point I realized that his recollection was faulty, since Minkowski gave an incorrect formula, as Siegel pointed out, and that was known to most experts. If I was asking questions on what he did in his twenties or thirties, he might have remembered things better, but at that time I did not take into account the fact that he worked on the Siegel formula in his fifties.

I asked him whether he was writing something on a historical topic. He said, "I cannot write any more." To cheer him up I then said, "That's why I told you long ago to get a computer." He also said he was half blind. Toward the end of the meal he said, "I'd like to see the Riemann hypothesis settled before I die, but that is unlikely."

That reminded me of a party at Borel's place in the 1970s. Wei-Liang Chow was the guest of honor. I was talking with Chow and Borel about a passage in Charlie Chaplin's autobiography. In it Chaplin in his twenties met a fortuneteller in San Francisco, who told him that he would make a tremendous fortune, would be married so many times with so many children, and would die of bronchial pneumonia at the age of eighty-two. Hearing this story, Weil said, "Well, in my autobiography I might write that in my youth I was told by a fortuneteller that I would never be able to solve the Riemann hypothesis."

When we left the dining hall and were walking to the parking lot, he said, "You are certainly disappointed, but I am disappointed too," and added, after a few seconds, "with myself." He knew that I was expecting him to say something about Siegel's work. He again said, "I cannot write any more." I drove him home and left. He was able to walk slowly but I could not say he was in good shape; still he was not in terrible shape, and so I had a sense of relief. While driving home alone under still drizzling rain, I could not but recall our hotel encounter in 1955 and the lunch in 1957, though I did not think much about the possibility that I would never see him again.

André Weil as a mathematician will of course be remembered by his colossal accomplishments witnessed by the three volumes of his *Collected Papers* and several books, the trilogy mentioned at the beginning in particular. In my mind, however, he will remain chiefly as the figure with two mutually related characteristics: First, he was flexible and receptive of new ideas of others and new directions, quite unlike many of the younger people these days who can work only within a well-established framework. Second, more importantly and in a similar vein, he had deep and penetrating understanding of mathematics, or rather, he strived tirelessly to understand the real meaning of every basic mathematical phenomenon, and to present it in a clearer form and in a better perspective. He did so by endowing each subject with new concepts and setting up new frameworks, always in a fresh and fundamental way. In other words, he was not a mere problem-solver. Clearly his death marked the end of an era, and at the same time left a large vacuum which will not easily be filled for a long time to come.

## Endnotes

1) Each number in brackets refers to the article designated by that number in his *Collected Papers* with "19" omitted.

2) It seems that [35b] is the first paper which mentions the fact that the coordinate ring of a variety is integral over a subring obtained by considering suitable hyperplanes (see *Collected Papers*, vol. I, p. 89). Zariski attributed it to E. Noether. It is my impression that she considered generic hyperplane sections, but not the fact of elements being integral. Weil agreed with me on this and said: "Perhaps Zariski didn't like to refer to the work of a younger colleague, a common psychological phenomenon." On the other hand, though he must have had his own citation policy, frankly I had difficulty in accepting it occasionally. See 9) below.

3) As to my paper on "Reduction of algebraic varieties etc." he said "il (Shimura) me dit, il eût plutôt eu en vue d'autres applications." (*Collected Papers*, vol. II, p. 542.) This is not correct. Probably he misunderstood me when I told him that I was interested in Brauer's modular representations at one time. Brauer was also a participant of the conference.

4) Mautner was responsible for introducing Weil to Tamagawa's idea; see Weil's comments on [59a].

5) In his *Collected Papers* he says practically nothing about his second visit, though he mentions it; see vol. II, p. 551.

6) This is what he told me. In his autobiography, however, the story is assigned to an earlier period, which may be true.

7) This is also what he told me. A somewhat different version is given in his autobiography.

8) There is a big difference. In order to win a lottery, we have to buy a ticket, but by doing so, we put our trust in the fairness of the system.

9) Whenever he spoke of strong approximation in algebraic groups, he always referred to Kneser's theorem. That is so in [65], for example, which is understandable. But that was always so even in his lectures in the 1960s, though in [62b] Eichler is mentioned in connection with the fact that the spinor genus of an indefinite quadratic form consists of a single class. However strange it may sound, it is possible, and even likely, that he was unable to recognize Eichler's fundamental idea and decisive result on strong approximation for simple algebras and orthogonal groups, and he knew only its consequence about the spinor genus. In his *Collected Papers* he candidly admits his ignorance in his youth. Though he had wide knowledge, his ignorance of certain well-known facts, even in his late years, surprised me occasionally. He knew Hecke's papers to the extent he quoted them in his own papers. It would be wrong, however, to assume that he was familiar with most of Hecke's papers. Besides, his comments in his *Collected Papers* include many insignificant references. For these reasons, the reader of those comments may be warned of their incompleteness and partiality.

10) According to Weil, once Hasse in such a uniform visited Julia, who became anxious about the possibility that he would be viewed as a collaborator.

11) In [65] he says, "On a ainsi retrouvé, quelque peu généralisées, tous les résultats démontrés par Siegel au cours de ses travaux sur les formes quadratiques, ainsi que ceux énoncés à la fin de [12] (Siegel's Annalen paper in 1952) á l'exception des suivants. Tout d'abord, ... ." (Collected Papers, vol. III, p. 154). I think this is misleading, since the list of exceptions does not include the case of inhomogeneous forms, which Siegel investigated. It is

true that Siegel's product formula for an inhomogeneous form in general can be obtained from the "formule de Siegel" (in Weil's generalized form, combined with some nontrivial calculations of the Fourier coefficients of Eisenstein series), and, one might say, that is not so important. Still, it should be mentioned at least that the inhomogeneous case is not just the matter of the Tamagawa number, and that nobody has ever made such explicit calculations in general, even in the orthogonal case. In the mid 1980s I asked Weil about this point, but he just said, "I don't remember."

12) In [76c] he reviews the complete works of Eisenstein; also the title of [76a] is *Elliptic functions according to Eisenstein and Kronecker*. It is believable, however, that he did not study Eisenstein's papers on quadratic forms in detail, though he must have been aware of them.

*          *          *

There was an interesting sequel to the publication of the above article. In early 1998 the editor of the *Notices* in charge showed my manuscript to Sylvie Weil, the eldest daughter of André, who endorsed it. After that, however, there was something which made her unhappy. I believe that it was related to the choice of the photos that would appear in the Weil issue of the *Notices,* but I am not certain. In any case, in a letter dated December 19, 1998 she wrote me that she was upset by my description of her father's weakened condition on the occasion of my last lunch with him. She said, "I would be extremely grateful if, as a favour to me, you would shorten that description, or perhaps not include it at all." So I wrote back to her as follows.

December 28, 1998

Dear Ms. Weil:

Frankly I was quite surprised by your unhappiness with my description of the last lunch I had with your late father. I read that passage again and found nothing awkward. After all, Borel, Tony Knapp, and his wife Susan read my article and were not disturbed by it. To be certain, I asked one of my colleagues Hale Trotter and his wife Kay to offer their opinions on that portion, They were favorable.

Of course you are in a different position, and I understand your feeling. However, let me emphasize that this article is aimed at the general mathematical public, and I thought it was my duty to present various aspects of the life of your father, especially in connection with me, as vividly and faithfully as possible. It is unfortunate that you simply viewed that passage as the description of his poorer moments. That the above-mentioned people found it positively will at least make you feel differently, I hope. Let me now explain in more detail that there are several points you might have missed.

First of all, that part must be read in the context of the whole article, or at least in contrast with its first few pages or with what I wrote (or actually what he said) about old French mathematicians. Next, more importantly, it tells something about what his existence meant to me. In his old age, he was still interested in mathematics, and he was trying to be useful to me. For various reasons he was sad; so was I. It is that sad feeling that I wanted to convey, though I avoided using the word "sad" or "sadness." Perhaps you will understand better if you put yourself in the narrator's position.

There are also meaningful passages concerning the Riemann hypothesis, which are best presented in the context of our lunch conversation. I recorded them for posterity.

Clearly he wished he could have done better, but he was 90, and in a sense he did well that day. I knew some famous people who were pompous and pretentious even after the age of 80. Not André Weil. He was honest about himself; he didn't pretend anything. As to the sentence, "You are certainly disappointed, but I am disappointed too," "with myself": Who else could have uttered those sincere words? Lesser men can never speak in that fashion. Over all he behaved with dignity, and we all should be proud of him. The question is whether the reader will feel the same way, but the reaction of my friends make me think that will be the case.

I guess this letter alone may not be able to quell your unhappiness immediately, but hope that it will do so eventually. At least let me assure you that practically no reader will take that part negatively.

Sincerely yours,<br>
Goro Shimura

I don't think she was really unhappy with my article, as she did not complain when it was shown to her. I think her request to me was her way of venting her anger toward something for which she thought I was responsible.

Some time in 1998 *Gazette des Mathématiciens,* published by Société Mathématique de France, planned a special issue on Weil, and asked me to send the above article of mine to them, which I did. A few weeks later, in March 1998, I received the proofs, which I corrected and returned to them. Then in a letter dated March 29, 1999, the chief editor, Daniel Barsky, demanded the deletion of 36 lines describing Weil's conditions, by saying that those passages would certainly hurt the majority of French mathematicians. Interestingly, note No. 12 at the end of my article was included in those 36 lines. He said that he would not publish it without the suppression. It was obvious that Sylvie pressured him. In fact, I asked the opinion of a few of my French friends in Paris, and none of them were in favor of the suppression. I refused to delete those passages, and my article was not published in that journal.

# AFTERWORD

First of all, I must note that I have written a book in Japanese, titled *Kioku no Kiri-ezu,* whose contents are almost the same as those of the present book, and which will be published by Chikuma Shobo in Tokyo. Thus there are two books with much in common, but one is not a translation of the other. I wrote both at the same time, using expressions in both languages comfortable for me, which naturally caused some differences, if not contradictions. There are also some passages in one book, that have no corresponding sections in the other. As a general rule, the present volume has more material concerning mathematics and mathematicians than the Japanese version.

My mathematical works, excluding those in book form and a few minor articles, are compiled in

*Collected Papers,* vol. I–IV, Springer, 2003.

This has a list of all the articles up to 2001, including those omitted in the collection. My work after 2001 can be checked on the Web.

To obtain many basic pieces of information concerning historical events, I was greatly helped by the Gest Oriental Library of Princeton University and the editorial department of Chikuma Shobo. It is my great pleasure to express my deepest thanks to them.

Printed in the United States of America